Simulation of Groundwater and Surface-Water Resources and Evaluation of Water-Management Alternatives for the Chamokane Creek Basin, Stevens County, Washington

By D. Matthew Ely and Sue C. Kahle

Prepared in cooperation with the United States Bureau of Indian Affairs and the Washington State Department of Ecology

Scientific Investigations Report 2012–5224

U.S. Department of the Interior
U.S. Geological Survey

U.S. Department of the Interior
KEN SALAZAR, Secretary

U.S. Geological Survey
Marcia K. McNutt, Director

U.S. Geological Survey, Reston, Virginia: 2012

For more information on the USGS—the Federal source for science about the Earth, its natural and living resources, natural hazards, and the environment, visit http://www.usgs.gov or call 1–888–ASK–USGS.

For an overview of USGS information products, including maps, imagery, and publications, visit http://www.usgs.gov/pubprod

To order this and other USGS information products, visit http://store.usgs.gov

Any use of trade, product, or firm names is for descriptive purposes only and does not imply endorsement by the U.S. Government.

Although this report is in the public domain, permission must be secured from the individual copyright owners to reproduce any copyrighted materials contained within this report.

Suggested citation:
Ely, D.M., and Kahle, S.C., 2012, Simulation of groundwater and surface-water resources and evaluation of water-management alternatives for the Chamokane Creek basin, Stevens County, Washington: U.S. Geological Survey Scientific Investigations Report 2012–5224, 74 p.

Contents

Contents—Continued

Figures

Figures—Continued

Figures—Continued

Tables

Conversion Factors and Datums

Inch/Pound to SI

Multiply	By	To obtain
Length		
inch (in.)	2.54	centimeter (cm)
inch (in.)	25.4	millimeter (mm)
foot (ft)	0.3048	meter (m)
mile (mi)	1.609	kilometer (km)
Area		
square mile (mi^2)	259.0	hectare (ha)
square mile (mi^2)	2.590	square kilometer (km^2)
Volume		
gallon (gal)	3.785	liter (L)
gallon (gal)	0.003785	cubic meter (m^3)
gallon (gal)	3.785	cubic decimeter (dm^3)
million gallons (Mgal)	3,785	cubic meter (m^3)
cubic foot (ft^3)	28.32	cubic decimeter (dm^3)
cubic foot (ft^3)	0.02832	cubic meter (m^3)
cubic mile (mi^3)	4.168	cubic kilometer (km^3)
acre-foot (acre-ft)	1,233	cubic meter (m^3)
acre-foot (acre-ft)	0.001233	cubic hectometer (hm^3)
Flow rate		
foot per day (ft/d)	0.3048	meter per day (m/d)
cubic foot per second (ft^3/s)	0.02832	cubic meter per second (m^3/s)
gallon per day (gal/d)	0.003785	cubic meter per day (m^3/d)
million gallons per day (Mgal/d)	0.04381	cubic meter per second (m^3/s)
Hydraulic conductivity		
foot per day (ft/d)	0.3048	meter per day (m/d)
Hydraulic gradient		
foot per mile (ft/mi)	0.1894	meter per kilometer (m/km)
Transmissivity*		
foot squared per day (ft^2/d)	0.09290	meter squared per day (m^2/d)

Temperature in degrees Celsius (°C) may be converted to degrees Fahrenheit (°F) as follows:

$$°F=(1.8×°C)+32.$$

Temperature in degrees Fahrenheit (°F) may be converted to degrees Celsius (°C) as follows:

$$°C=(°F-32)/1.8.$$

Datums

Vertical coordinate information is the North American Vertical Datum of 1988 (NAVD 88).

Horizontal coordinate information is referenced to the North American Datum of 1983 (NAD 83).

Altitude, as used in this report, refers to distance above the vertical datum.

*Transmissivity: The standard unit for transmissivity is cubic foot per day per square foot times foot of aquifer thickness [(ft^3/d)/ft^2]ft. In this report, the mathematically reduced form, foot squared per day (ft^2/d), is used for convenience.

Simulation of Groundwater and Surface-Water Resources and Evaluation of Water-Management Alternatives for the Chamokane Creek Basin, Stevens County, Washington

By D. Matthew Ely and Sue C. Kahle

Abstract

A three-dimensional, transient numerical model of groundwater and surface-water flow was constructed for Chamokane Creek basin to better understand the groundwater-flow system and its relation to surface-water resources. The model described in this report can be used as a tool by water-management agencies and other stakeholders to quantitatively evaluate the effects of potential increases in groundwater pumping on groundwater and surface-water resources in the basin.

The Chamokane Creek model was constructed using the U.S. Geological Survey (USGS) integrated model, GSFLOW. GSFLOW was developed to simulate coupled groundwater and surface-water resources. The model uses 1,000-foot grid cells that subdivide the model domain by 102 rows and 106 columns. Six hydrogeologic units in the model are represented using eight model layers. Daily precipitation and temperature were spatially distributed and subsequent groundwater recharge was computed within GSFLOW. Streamflows in Chamokane Creek and its major tributaries are simulated in the model by routing streamflow within a stream network that is coupled to the groundwater-flow system. Groundwater pumpage and surface-water diversions and returns specified in the model were derived from monthly and annual pumpage values previously estimated from another component of this study and new data reported by study partners.

The model simulation period is water years 1980–2010 (October 1, 1979, to September 30, 2010), but the model was calibrated to the transient conditions for water years 1999–2010 (October 1, 1998, to September 30, 2010). Calibration was completed by using traditional trial-and-error methods and automated parameter-estimation techniques. The model adequately reproduces the measured time-series groundwater levels and daily streamflows. At well observation points, the mean difference between simulated and measured hydraulic heads is 7 feet with a root-mean-square error divided by the total difference in water levels of 4.7 percent. Simulated streamflow was compared to measured streamflow at the USGS streamflow-gaging station—Chamokane Creek below Falls, near Long Lake (12433200). Annual differences between measured and simulated streamflow for the site ranged from -63 to 22 percent. Calibrated model output includes a 31-year estimate of monthly water budget components for the hydrologic system.

Five model applications (scenarios) were completed to obtain a better understanding of the relation between groundwater pumping and surface-water resources. The calibrated transient model was used to evaluate: (1) the connection between the upper- and middle-basin groundwater systems, (2) the effect of surface-water and groundwater uses in the middle basin, (3) the cumulative impacts of claims registry use and permit-exempt wells on Chamokane Creek streamflow, (4) the frequency of regulation due to impacted streamflow, and (5) the levels of domestic and stockwater use that can be regulated. The simulation results indicated that streamflow is affected by existing groundwater pumping in the upper and middle basins. Simulated water-management scenarios show streamflow increased relative to historical conditions as groundwater and surface-water withdrawals decreased.

Introduction

Chamokane Creek basin is a 179 mi² area that borders and partially overlaps the Spokane Indian Reservation in southern Stevens County in northeastern Washington (fig. 1). The basin is a roughly boot-shaped, northwest-to-south-trending basin about 28 mi long and 7 mi wide. Chamokane Creek flows toward the east through the Camas Valley to near the town of Springdale, Washington, where the creek turns southeast and then flows generally south through the Chamokane Valley and Walkers Prairie toward the town of Ford, Washington. Mean September streamflow in Chamokane Creek at Chamokane Falls, as recorded at U.S. Geological Survey (USGS) streamflow-gaging station 12433200 (fig. 1), for 1971–2008 was 27 ft³/s.

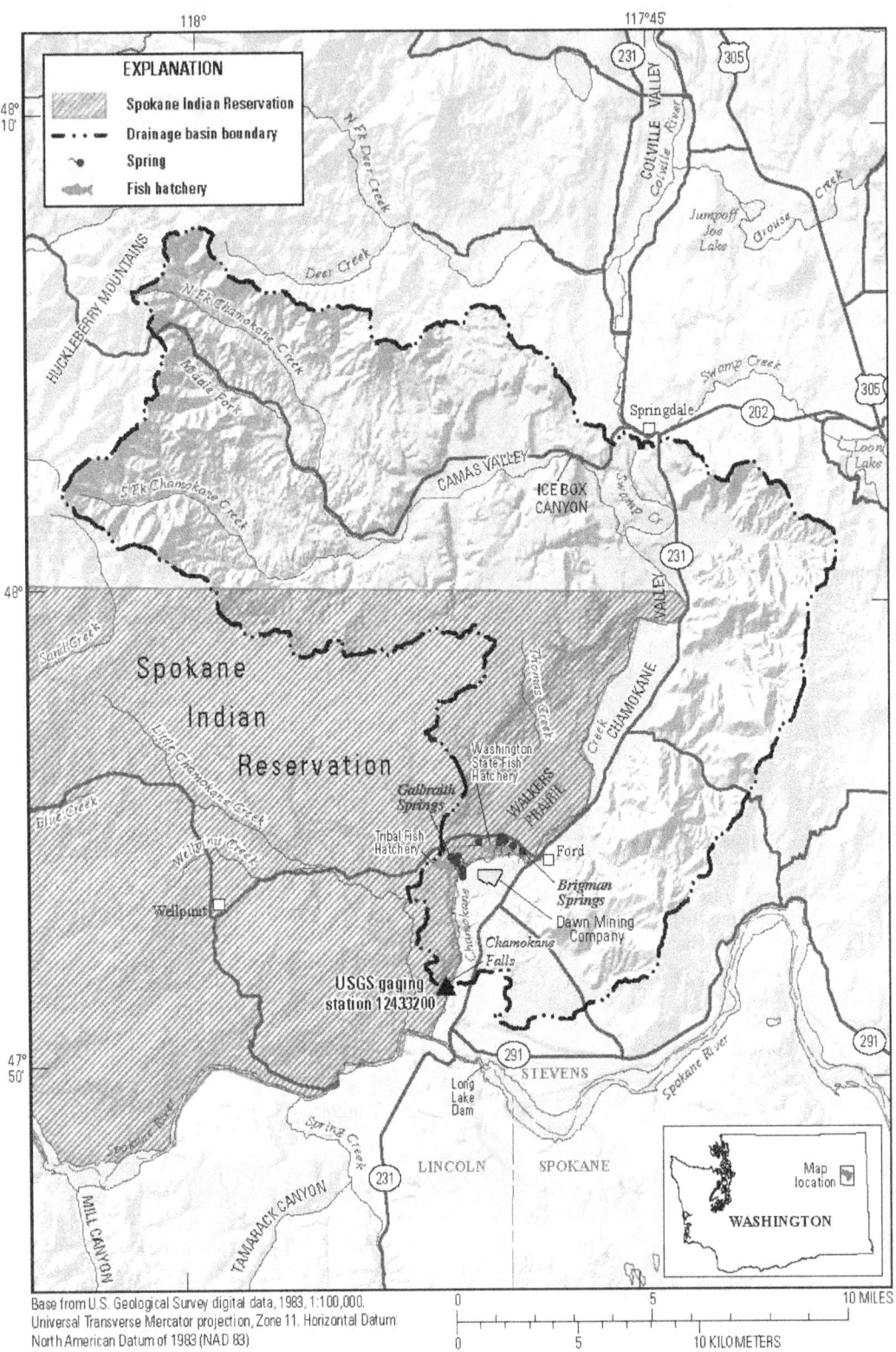

Figure 1.   Location of the Chamokane Creek basin, Stevens County, Washington.

In 1979, most of the water rights in the Chamokane Creek basin were adjudicated by the United States District Court. Since the 1979 adjudication, the District Court has issued various amendments and orders that affect water users in the basin. The Chamokane Creek Adjudication requires that junior water right holders on Chamokane Creek, and its tributaries, be regulated in favor of the more senior water right of the Spokane Tribe. These senior water rights were granted as reserved water rights for irrigation and protection of the Spokane Tribe's fishing rights in Chamokane Creek. A court-appointed Water Master regulates junior water rights when the mean daily 7-day low flow (mean daily discharge computed over 7 consecutive days) becomes less than 24 ft^3/s (27 ft^3/s for rights issued after December 1988) at Chamokane Falls, as recorded at U.S. Geological Survey (USGS) streamflow-gaging station 12433200 (fig. 1). Regulation has been necessary in 3 recent years (2001, 2005, and 2009). The non-Reservation areas of the basin are closed to any additional groundwater or surface-water appropriation, with the exception of "permit exempt" uses of groundwater. These exempt uses in Washington State do not require a water right and include stock watering, lawn or non-commercial garden watering, single or group domestic uses of as much as 5,000 gal/d, and small-scale industrial use not to exceed 5,000 gal/d (Washington State Department of Ecology, 2006).

The 1979 District Court Judgment contends that groundwater pumping in the Upper Chamokane region had no effect on the flow of Chamokane Creek because groundwater in the upper region was considered to be part of a separate aquifer from that in the Chamokane Valley. Despite the ruling, there are concerns about the effects of future groundwater development that may occur in the upstream end of the basin, particularly outside the reservation boundary. The relation between the upper (Camas Valley and Chamokane Creek headwaters) and middle (Chamokane Valley) Chamokane Creek basin groundwater systems has not been directly studied. With increasing population and residential development, permit-exempt groundwater use is expected to continue to grow, and the potential effects of this growth on Chamokane Creek are unknown.

To evaluate these concerns, the USGS began a two-phase study in 2007 with the primary goals of describing the groundwater and surface-water system of the valley-fill deposits of the Chamokane Creek basin and assessing the effects of potential increases in groundwater pumping on groundwater and surface-water resources. The results from Phase 1 of this study are presented in Kahle and others (2010). The Phase 1 report includes descriptions of the hydrogeologic setting and groundwater/surface-water interactions in the basin, and selected hydrologic datasets to support construction and calibration of a coupled groundwater and surface-water flow model in Phase 2.

This report describes Phase 2 of the study— the construction and calibration of a numerical flow model designed to describe groundwater/surface-water interactions in the basin. The model was used to provide a comprehensive water budget for the study area and evaluation of the possible regional effects of different water-management scenarios on the surface-water flow system. Model results specifically address a series of factual questions filed by the U.S. District Court.

Purpose and Scope

This report presents the results of Phase 2 of the study of the groundwater and surface-water resources in the Chamokane Creek basin. This report describes the hydrogeologic framework of the basin; model boundaries; water budgets; the construction, calibration, and application of the model; results of various water-management scenarios evaluated with the model; and model limitations. The purpose for constructing the model was to improve understanding of the groundwater-flow system and its relation to surface-water resources. The model development is presented and described, and includes information on the delineation of basin physical characteristics, distribution of daily climate data, spatial and temporal discretization of the aquifer system, boundary conditions, stresses, and hydraulic properties of the hydrogeologic units constituting the aquifer system. The results from the application of the model for five water-management scenarios also are presented, and the results are described relative to differences in simulated streamflow from the calibrated model.

Description of Study Area

Altitudes in the study area range from 1,420 ft near the outlet of the basin to 4,600 ft in the Huckleberry Mountains where Chamokane Creek has its headwaters. The creek flows east from its headwaters into Camas Valley, a southwest-northeast oriented valley about 6.5 mi long (fig. 1). From Camas Valley, Chamokane Creek flows through Ice Box Canyon and then southeast about 4 mi where it is slightly entrenched in a system of outwash terraces. The creek continues southeast to a bedrock outcrop where it changes course and flows southwest into Walkers Prairie. The three headwater branches, the North Fork Chamokane Creek, the Middle Fork Chamokane Creek, and the South Fork Chamokane Creek, which form the mainstem of Chamokane Creek, are all intermittent in their upper reaches. Chamokane Creek becomes perennial most years in Camas Valley and remains so until it reaches the Walkers Prairie area where it again becomes intermittent due to seasonal infiltration losses through the channel bed.

On the southern end of Walkers Prairie, a series of large springs discharge from an east-west oriented bluff west of Ford (fig. 1). Discharge from these springs and outflow from two hatcheries provides most of the perennial flow in Chamokane Creek from near Ford to the confluence with the Spokane River. Chamokane Falls is about 1.5 mi upstream of the confluence where Chamokane Creek flows over a bedrock outcrop (fig. 1).

A low-altitude drainage divide near Springdale, Washington, between the north-flowing Colville River and the south-flowing Chamokane Creek, is underlain by glacial outwash and till associated with the Colville sublobe of the Cordilleran ice sheet and, at greater depths, by thick clay and silt deposited in large Pleistocene lakes (Kahle and others, 2003, 2010). These unconsolidated deposits form a shallow surface-drainage divide in an otherwise broad and continuous Colville-Chamokane Valley. A pre-glacial Columbia River may have flowed southward through the present-day Colville and Chamokane Valley, resulting in the long and wide valley that is visible today (Willis, 1887; Wozniewicz, 1989; Carrara and others, 1996).

Most mountainous areas in the basin are covered with pine, fir, and larch forests that are the basis for the historical and 2010 lumber industry in the area. In the lowland areas of the basin, agricultural land use is widespread, including grazing and hay production, along with scattered developed areas, including the town of Ford, Washington. About 34 mi^2 of the Spokane Indian Reservation lies within Chamokane Creek basin with Chamokane Creek's east bank forming the eastern border of the Reservation from just north of 48 degrees latitude to the confluence of Chamokane Creek and the Spokane River (fig. 1). Two fish hatcheries, one operated by the State and one by the Spokane Tribe, are on the northern side of Chamokane Creek west of Ford.

The climate in the study area varies from subhumid to semiarid and is characterized by warm, dry summers and cool, moist winters (Molenaar, 1988). Mean annual (1923–2007) precipitation for the nearest long-term weather station in Wellpinit, Washington, is 18.95 in. (Western Region Climate Center, 2010). Historically, most precipitation falls as snow during the 5-month period from November through March. Average annual precipitation for 1971–2000 ranges from a minimum of about 14 in. at the southern edge to more than 25 in. in the northernmost headwaters of the basin (Western Regional Climate Center, 2010).

Hydrogeology

Geologic Setting

The oldest rocks in the Chamokane Creek basin occur throughout the Huckleberry Mountains and are composed mostly of argillite, a weakly metamorphosed shale or mudstone. These rocks are about 1.5 billion years old and as such are some of the oldest rocks in Washington State. Somewhat younger rocks including quartzite and limestone have small surface exposures southeast of Springdale and immediately southwest of the confluence of Swamp and Chamokane Creeks (Stoffel and others, 1991). Granitic rocks, about 100 million years old, intruded the older rocks and are now exposed at land surface along the lower reaches of the

Middle and South Fork Chamokane Creek, along the eastern side of the Chamokane Creek basin from near Springdale to Happy Hill, and at Chamokane Falls. Basalt that is about 16 million years old mantles older rocks in parts of the basin and covers much of the west-central part of the study area, forming the bluff on the western side of Walkers Prairie and the perimeter of Camas Valley.

During the Pleistocene (2.6 million years ago to 11,000 years ago), the study area was subjected repeatedly to the erosional and depositional processes associated with glacial and interglacial periods resulting in an assemblage of unconsolidated sediment that overlies much of the bedrock in the study area and is thickest along the valley floors (Wozniewicz, 1989; Kahle and others, 2003, 2010). Effects of the most recent glaciation (about 16,000–11,000 years ago) are the most easily recognized in the study area. The southernmost limit of the Colville sublobe is marked by a well-developed moraine near the town of Springdale (Carrara and others, 1996), where hummocky topography resulted from the deposition of material pushed along the ice front and from melting of sediment laden ice. Melt water from the Colville sublobe created a large outwash plain and series of gravel terraces extending from near Springdale to south of Ford. Just beyond the southern limit of the sublobe are former glacial meltwater channels now occupied by Swamp Creek.

Glacial Lake Missoula was created in the ancestral Clark Fork in northwestern Montana when the Purcell Trench lobe blocked westward drainage of glacial meltwater in northern Idaho. Enormous catastrophic floods occurred over a 2,000-year period when the ice dam of the Purcell Trench lobe periodically failed, sending floodwaters west and southwest throughout parts of Montana, Idaho, Washington, and Oregon, before eventually reaching the Pacific Ocean. Some of the earlier Missoula floods, following one of the more northern floodways, left behind giant current dunes north of Loon Lake before exiting westward through the Sheep Creek spillway into the Colville Valley near Springdale and then southward through the Chamokane Valley (Kiver and Stradling, 1982; Carrara and others, 1995). Meltwater from the Colville sublobe later reworked the earlier flood deposits.

Glacial Lake Columbia, impounded by the Okanogan lobe near present day Grand Coulee, was the largest glacial lake in the path of the Missoula floods. This lake was long-lived (2,000–3,000 years) and had a typical surface altitude of 1,640 ft; however, the altitude reached 2,350 ft during maximum blockage by the Okanogan lobe (Atwater, 1986). The higher surface altitude of Glacial Lake Columbia probably occurred early, whereas the lower and more typical surface altitude of the lake occurred in later glacial time (Richmond and others, 1965; Waitt and Thorson, 1983; and Atwater, 1986). At the lower surface altitude (1,640 ft), Glacial Lake Columbia extended into the Chamokane Valley to near Ford and east to the Spokane area, where clayey lake sediment is intercalated with Missoula flood sediment. At the higher surface altitude of Glacial Lake Columbia (2,350 ft), the glacial lake would have flooded the entire combined Colville-Chamokane Valley, nearly reaching the top of the basalt bluffs on the western edge of Walkers Prairie.

As the Colville sublobe neared its southern maximum, the outflow of glacial meltwater draining Camas Valley was pushed southward presumably by the front of the glacial ice and the morainal material causing it to incise a narrow channel (Ice Box Canyon) through a northern limb of Lyons Hill (McLucas, 1980). When the Colville sublobe reached its southernmost position near Springdale, it may have blocked Camas Valley to the west creating a local glacial lake (Wozniewicz, 1989). Thick fine-grained sediment (clay and silt) that occur at depth in the Camas Valley are likely associated with a local glacial lake and with Glacial Lake Columbia (Kahle and others, 2010).

The surficial geology of the Chamokane Creek basin consists of the following ten geologic units summarized in Kahle and others (2010).

Alluvial deposits (Qal): This unit includes channel and overbank deposits of Chamokane and Swamp Creeks and alluvial-fan deposits at the mouths of streams tributary to Chamokane Creek. The unit consists mostly of stratified silt and sand with some gravel and minor amounts of clay deposited by flowing water and generally is from 1 to 30 ft thick.

Glacial outburst flood deposits (Qf): This unit includes glacial-outburst flood deposits that consist of sand with sparse pebbles, cobbles, and boulders deposited by catastrophic draining of Glacial Lake Missoula into lower energy environments along the margins of the floods. These deposits are mapped in the southern part of basin. This unit is as much as 100 ft thick within the Chamokane Creek basin.

Glaciofluvial deposits (Qgf): This unit includes mostly stratified and well-sorted sand, gravel, and cobble outwash deposited by glacial meltwater from the Colville sublobe. It also includes outwash deposits along the northern part of the Camas Valley and outwash and reworked outburst flood deposits along the Chamokane Valley floor and Walkers Prairie. Although most of the Qgf is coarse-grained outwash, lenses of silt, clay, and till occur locally. The thickness of the unit generally is from 20 to 200 ft.

Glacial deposits, undifferentiated (Qgu): This unit includes a heterogeneous mixture of silt, sand, gravel, and clay that may include loess deposits and older glacial till. The unit mantles the upland areas in the west central part of the study area and generally is from 5 to 20 ft thick.

Glacial till (Qti): This unit includes mostly unsorted and unstratified clay, silt, sand, and gravel deposited by the Colville sublobe. Near Springdale, the unit includes the terminal moraine of the Colville sublobe. Locally, the unit contains stratified sand and gravel and generally is from 10 to 80 ft thick.

Glaciolacustrine deposits (Qla): This unit includes mostly clay and silt lake sediments deposited in ice-marginal lakes. The unit underlies the Chamokane Creek Valley and is overlain by younger deposits within the basin. Just north of the Chamokane Creek basin, about 2 mi northeast of Springdale, the unit occurs at land surface in the Colville Valley floor. The unit includes thin and discontinuous beds of sand and gravel in places. Along the axis of the Chamokane Valley, the thickness of the unit is commonly about 300 ft.

Loess (Qlo): This unit includes silt and fine sand, and minor amounts of clay, deposited by winds. The unit forms an extensive blanket on the basalt upland of the Lyons Hill area and has a limited extent to the east of Happy Hill in the southeastern part of the study area. Thickness of the unit typically ranges from 1 to 15 ft.

Mass-wasting deposits (Qmw): This unit includes poorly sorted angular rock fragments deposited as talus at the base of steep slopes and heterogeneous mixtures of unconsolidated surficial material and rock fragments deposited by landslides. The largest surface exposures of this unit are along the basalt bluff on the western edge of Walkers Prairie and on the western side of Happy Hill near Ford. Thickness of the unit varies, but may exceed 200 ft in places.

Basalt (Miocene) (Mb): This unit includes the Grande Ronde Basalt of the Columbia River Basalt Group, a dense, dark basalt with fine to coarse interbeds. Interbeds may be part of the Latah Formation, which was deposited along the margins of the basalt flows in eastern Washington. Thickness of the unit in the study area is uncertain, but may be more than 500 ft.

Bedrock (Tertiary to Middle Proterozoic) (Tybr): This unit includes sedimentary, metasedimentary, and intrusive and extrusive igneous rocks. Specific rock types include shale, conglomerate, dolomite, limestone, argillite, gneiss, schist, slate, quartzite, and granite. The unit is exposed in much of the high-altitude areas of the basin. The depth to bedrock in the Chamokane Creek Valley beneath the unconsolidated sediments is largely unknown and likely varies considerably. Based on information from wells used in this study, the depth to bedrock along most of the central part of the Chamokane Creek Valley may be as much as 600 ft.

Hydrogeologic Units

The geologic units described previously were grouped into six hydrogeologic units based on similar lithologic characteristics and large-scale hydrologic properties (Kahle and others, 2010). The six hydrogeologic units described in this report include the Upper outwash aquifer, the Landslide unit, the Valley confining unit, the Lower aquifer, the Basalt unit, and the Bedrock unit. Lithologic and hydrologic characteristics of these units are summarized in figure 2. The mapped extent of the four units that occur at land surface in the study area is shown in figure 3. The subsurface extent of the units is illustrated on two hydrogeologic sections (fig. 4).

Hydrogeologic unit	Unit label	Range of thickness [estimated average thickness], in feet	Lithologic and hydrologic characteristics
Upper outwash aquifer	UA	2 – 280 [80]	Unconfined aquifer consisting of gravel, cobbles, boulders, and sand with minor silt and or clay interbeds. Unit occurs in the Chamokane Valley from Swamp Creek southward throughout Walkers Prairie. UA occurs in Camas Valley, but is thinner and less productive. Includes glaciofluvial, glacial outburst flood, and alluvial deposits. Near Springdale, UA includes till, a lower-permeability deposit that includes compacted and poorly sorted silt, sand, gravel, and cobbles with lenses of moderately sand and gravel.
Landslide unit	LU	0 – 205 [150]	Poorly sorted deposits of broken basalt and sedimentary interbeds of the Columbia River Basalt Group, covered in places by glacial deposits. Unit occurs along the eastern slopes of the basalt mesa on the eastern uplands of the Spokane Indian Reservation and on the western flanks of Happy Hill near Ford. Unit likely is in hydraulic connnection with the UA in Walker's Prairie. Locally an aquifer with variable yields.
Valley confining unit	VC	4 – 480 [170]	Low-permeability unit consisting mostly of glaciolacustrine silt and clay. Unit occurs at depth throughout the Camas and Chamokane Valleys and is continuous northward into the Colville Valley. Discontinuous lenses of aquifer material within the unit contribute usable quantities of water to some wells, particularly in the southern part of the basin where Missoula flood deposits are interbedded with the glaciolacustrine deposits.
Lower aquifer	LA	3 – 170	Confined aquifer consisting of sand and some gravel. Unit occurs at depth in the Camas and Chamokane Valleys below the Valley confining unit and is continuous northward into the Colville Valley. Thickness and extent of unit is not well known.
Basalt unit	BT	40 – 500	Unit is composed of Columbia River Basalt, a dense, dark basalt with generally fine grained interbeds. Coarse-grained interbeds occur near Springdale. Water is contained in cracks and fractures and from zones between lava flows. Occurs in the eastern uplands of the Spokane Indian Reservation and on Lyons Hill. Unit is overlain in places by thin and discontinuous loess and/or glacial deposits.
Bedrock unit	BR	Not applicable	Unit includes argillite, conglomerate, dolomite, gneiss, schist, slate, quartzite, shale, limestone, and granite. Yields generally are small; numerous abandoned wells. Unit is overlain in places by thin and discontinuous alluvium, loess, and/or glacial fluvial deposits.

Figure 2.   Lithologic and hydrologic characteristics of the hydrogeologic units in the Chamokane Creek basin, Washington (modified from Kahle and others, 2010).

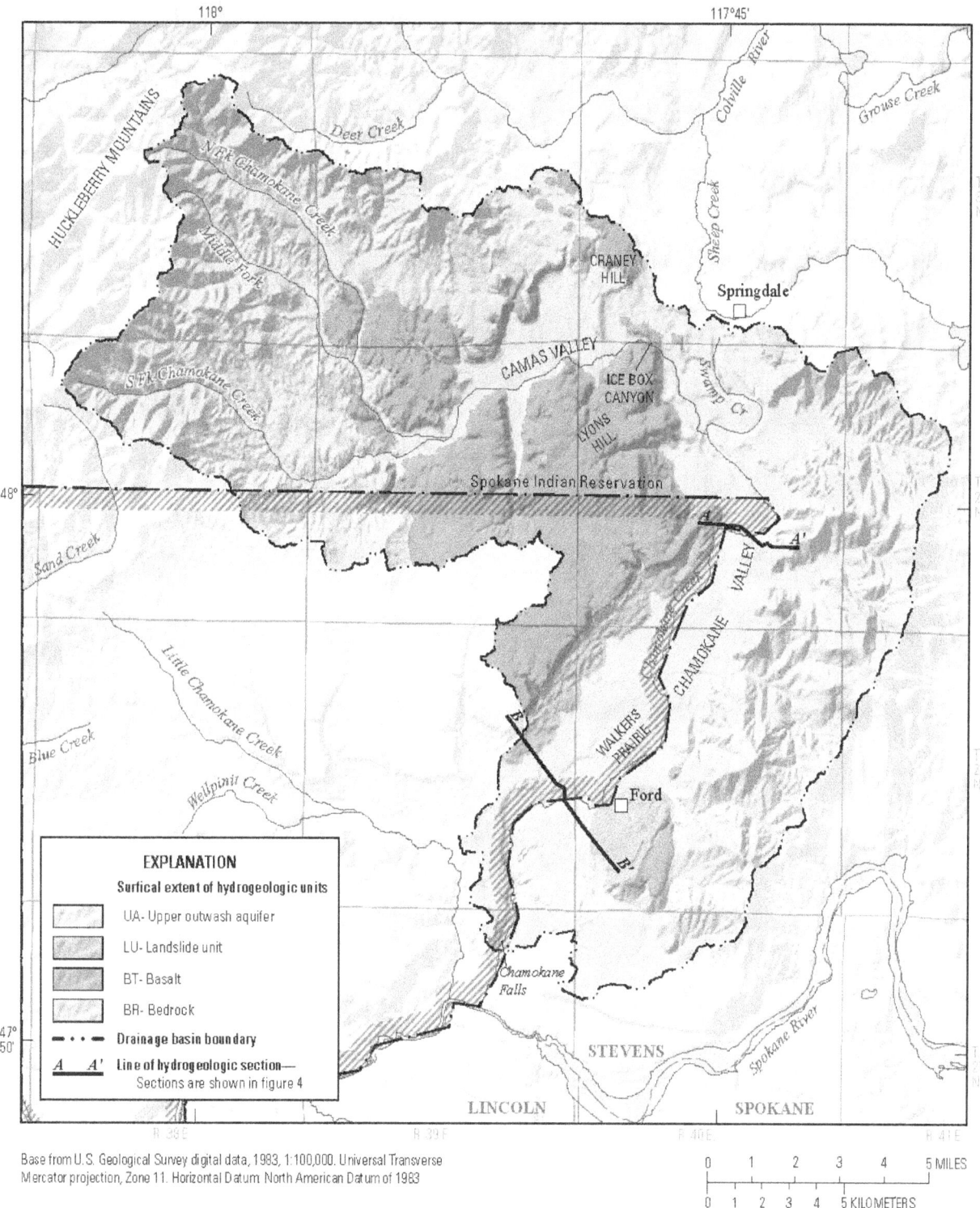

Figure 3.   Surficial extent of the hydrogeologic units in the Chamokane Creek basin, Washington (modified from Kahle and others, 2010).

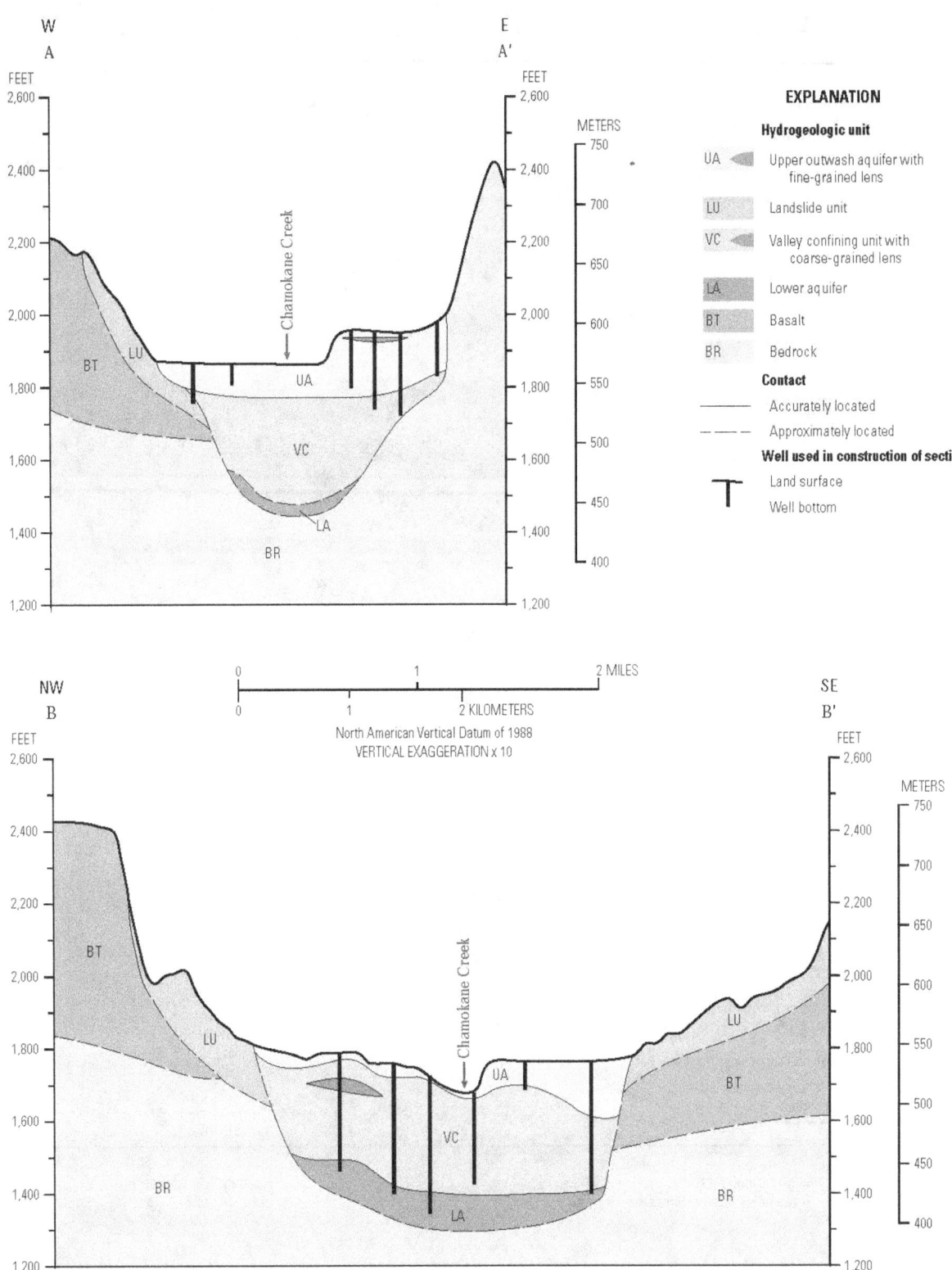

Figure 4. Hydrogeologic units of Chamokane Creek basin, Washington (modified from Kahle and others, 2010).

The Upper outwash aquifer (UA) of the Chamokane Creek basin is an unconfined aquifer consisting of sand, gravel, cobbles, and boulders, with minor silt and (or) clay interbeds. Previously described geologic units that constitute the bulk of this aquifer include glaciofluvial deposits (outwash), glacial outburst flood deposits, and alluvial deposits. Near Springdale, the aquifer includes till, a lower permeability deposit comprising compacted and poorly sorted silt, sand, gravel, and cobbles with lenses of moderately sorted sand and gravel.

The Upper outwash aquifer occurs along the length of the Chamokane and Camas Valleys (fig. 5). Although the Upper outwash aquifer exists over much of the valley floors, some places within the aquifer are not sufficiently saturated to yield sustainable quantities of water to wells (fig. 5). Wells are drilled into deeper units in these areas. For example, most wells south of Ford are drilled into the Lower aquifer because the Upper outwash aquifer in those areas generally yields insufficient quantities of water.

Aquifer thickness ranges from less than 50 ft along its margins to more than 150 ft where glacial terraces comprise the unit. From the Swamp Creek area southward through Walkers Prairie to near the mouth of the basin, the unit is extremely coarse-grained in most of the upper 20–30 ft where cobbles, boulders, and gravel are common. The lower part of the aquifer generally is composed of gravel and sand with few cobbles or boulders. In the Camas Valley, where the Upper outwash aquifer is composed mostly of recent alluvial deposits along stream courses, it contains more sand and clay, and is therefore less productive than along Walkers Prairie. The aquifer is about 15–30 ft thick over much of the Camas Valley, but thickens to about 80 ft near the outlet of the valley.

The Landslide unit (LU) is composed of poorly sorted deposits of broken basalt and sedimentary interbeds of the Columbia River Basalt Group, covered in places by glacial deposits. Locally, the Landslide unit is an aquifer with variable yields. The unit occurs along the eastern slopes of the basalt mesa on the eastern uplands of the Spokane Indian Reservation and on the western flanks of Happy Hill near Ford. The extent of this hydrogeologic unit at land surface is shown in figure 3. The approximate extent of this unit below land surface is shown in hydrogeologic sections A–A′ and B–B′ (fig. 4). As illustrated on the sections, the Landslide unit likely is in hydraulic connection with the Upper outwash aquifer along Walkers Prairie.

The Valley confining unit (VC) consists mostly of low-permeability glaciolacustrine silt and clay. The unit occurs at depth throughout the Camas and Chamokane Valleys and is continuous northward into the Colville Valley (Kahle and others, 2003, 2010). Discontinuous lenses of sand or gravel within the unit contribute usable quantities of water to some wells. Coarse-grained lenses within the Valley confining unit appear to be more common in the southern part of the basin where Missoula flood deposits following the Spokane River drainage are interbedded with the glaciolacustrine deposits. Thickness of the Valley confining unit commonly is 150–300 ft along the Camas and Chamokane Valleys and exceeds 300 ft in the central Camas Valley, Swamp Creek area, and near Ford (Kahle and others, 2010).

The Lower aquifer (LA) is a confined aquifer consisting of sand and some gravel that occurs at depth in the Camas and Chamokane Valleys beneath the Valley confining unit (figs. 4 and 6). The Lower aquifer is continuous northward into the Colville Valley (Kahle and others, 2003, 2010). The approximate southern extent of the Lower aquifer is shown as being truncated near Chamokane Falls based on outcrop of bedrock in that vicinity and well records that indicate the absence of a lower aquifer. The Chamokane basin Lower aquifer does not appear to be connected with similar deposits along the Spokane River. The thickness of the Lower aquifer was interpreted to be 125 and 138 ft, respectively, at two project wells that fully penetrate the unit and reach the underlying Bedrock unit (Kahle and others, 2010).

The Basalt unit (BT) is composed of Columbia River Basalt, a dense, dark basalt with generally fine-grained interbeds. Locally, coarse-grained interbeds occur in the unit west of Springdale. Water from cracks and fractures in the basalt and from zones between lava flows can supply usable quantities of water to wells. The Basalt unit occurs at land surface in the eastern uplands of the Spokane Indian Reservation and on Lyons Hill and Craney Hill (fig. 3). This hydrogeologic unit includes Miocene basalt overlain by thin and discontinuous Quaternary loess and glacial deposits.

The Bedrock unit (BR) includes argillite, conglomerate, dolomite, gneiss, schist, slate, quartzite, shale, limestone, and granite. The unit locally yields usable quantities of water where rocks are fractured, but yields generally are small and numerous abandoned wells occur in the unit. The Bedrock hydrogeologic unit includes geologic unit Tybr and thin and discontinuous Qal, Qlo, and Qgf that overly the bedrock.

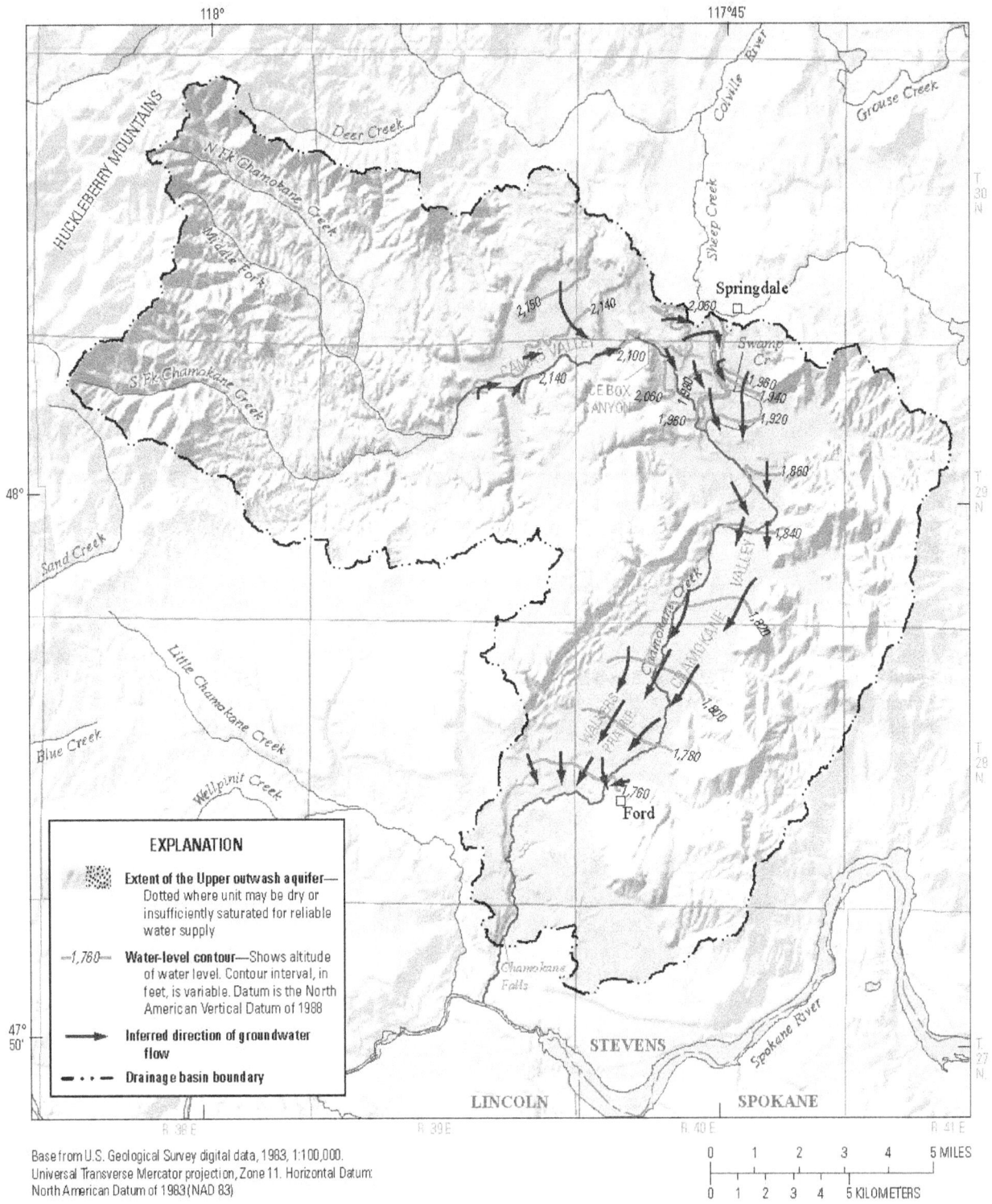

Figure 5. Areal extent, water-level altitudes, and inferred directions of groundwater flow in the Upper outwash aquifer in the Chamokane Creek basin, Washington (modified from Kahle and others, 2010).

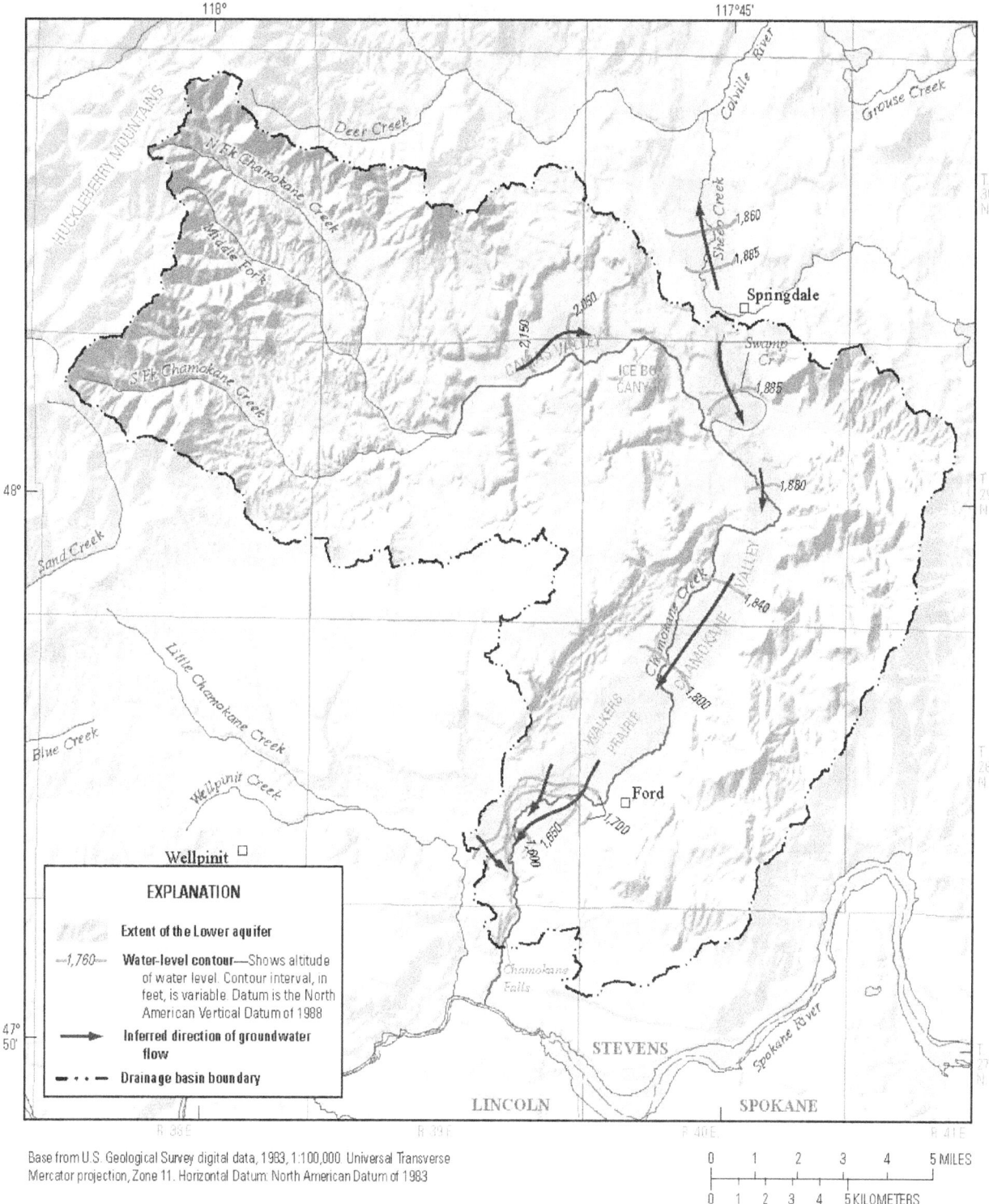

Figure 6.   Approximate extent, water-level altitudes, and inferred directions of groundwater flow in the Lower aquifer in the Chamokane Creek basin, Washington (from Kahle and others, 2010).

Hydraulic Characteristics

Horizontal hydraulic conductivity was estimated by Kahle and others (2010) and is summarized by hydrogeologic unit in the following table. Data were unavailable for the Landslide and Bedrock units. The median values of estimated hydraulic conductivities for the aquifers are similar in magnitude to values reported by Freeze and Cherry (1979) for similar materials—Upper outwash aquifer, 540 ft/d and Lower aquifer, 19 ft/d. The medians of estimated hydraulic conductivities for the Valley confining unit (10 ft/d), and the Basalt unit (3.7 ft/d) are higher than is typical for most of the material in these units because the data for confining units usually are from zones where lenses of coarse material exist and, in the case of the Basalt unit, where fractures or sedimentary interbeds exist. As a result, the data are biased toward the more productive zones in these units and are not representative of the entire unit. The minimum hydraulic conductivities for the hydrogeologic units illustrate that there are zones of low hydraulic conductivity in most units. Additionally, the range of hydraulic conductivities is at least three orders of magnitude for most units, indicating substantial heterogeneity and inherent uncertainty in estimating effective hydraulic conductivity for units represented in the model.

Hydrogeologic unit	Estimated hydraulic conductivity (feet per day)			
	Mini-mum	Median	Maxi-mum	Number of values
Upper outwash aquifer	15	540	7,900	10
Valley confining unit	2	10	860	4
Lower aquifer	4	19	3,000	8
Basalt unit	0.93	3.7	6.5	2

Estimates of horizontal hydraulic conductivity reported in several other investigations provide useful comparisons to the values estimated during this investigation. Buchanan and others (1988) reported an average horizontal hydraulic conductivity of 2,664 ft/d for the Upper outwash aquifer of Walkers Prairie based on a long-term aquifer test with multiple observation wells. On the Dawn millsite south of Ford, the horizontal hydraulic conductivity of the Upper outwash aquifer was estimated to be 14–140 ft/d based on aquifer test data (Washington State Department of Health, 1991). Golder Associates, Inc. (2008) reported an average horizontal hydraulic conductivity of 331 ft/d for the Lower aquifer near the southern end of Chamokane Creek basin based on specific capacity data. Horizontal hydraulic conductivity of the Lower aquifer near Galbraith Springs is approximately 1,300 ft/d, based on an aquifer test conducted by Rittenhouse-Zeman and Associates, Inc. (1989). Whiteman and others (1994) estimated the median hydraulic conductivity of the Grande Ronde Basalt over the Columbia Plateau as 4.9 ft/d based on specific capacity data.

Hydrologic System

Groundwater Recharge

Direct precipitation recharges the Upper outwash aquifer over its extent and streamflow recharges the aquifer where losing stream reaches directly overlie the aquifer. Significant mountain-front recharge also may occur along the perimeter of the aquifer where it is in contact with Landslide unit deposits or productive zones within the Basalt or Bedrock units.

Recharge to the Lower aquifer likely occurs in several areas. Water-level data indicate that recharge occurs from near Springdale through the Swamp Creek area where vertical head gradients between the Upper and Lower aquifers generally are downward (Kahle and others, 2010). Localized recharge also occurs along the walls of the Camas Valley and Walkers Prairie where coarse talus slopes and landslide deposits along basalt bluffs or glacial outwash fans overlie and interfinger with the otherwise continuous Valley confining unit.

Groundwater Movement

Horizontal groundwater flow in the Upper outwash aquifer moves from the topographically high tributary-basin areas toward the topographically lower valley floors. Water-level altitudes in the Upper outwash aquifer range from 2,150 ft in the Camas Valley to 1,760 ft near Ford (fig. 5). The general distribution of horizontal hydraulic gradients was about 13–50 ft/mi in the Camas Valley, about 80 ft/mi where Chamokane Creek exits Icebox Canyon to near its confluence with Swamp Creek, 20–30 ft/mi from south of Springdale through the Swamp Creek area, and 12–16 ft/mi along Walkers Prairie. The smallest gradient in the Upper outwash aquifer, about 12 ft/mi, was along Walkers Prairie.

Horizontal groundwater flow in the Lower aquifer is south to southwest from near Springdale to south of Ford (fig. 6). In the Camas Valley, horizontal groundwater flow is east to near the end of the valley where flow likely discharges into overlying sediments and Chamokane Creek near the end of the valley at the head of Icebox Canyon. Along the Chamokane Valley floor, water-level altitudes within the Lower aquifer range from 1,885 ft near Swamp Creek to 1,600 ft near the lower end of the basin. Horizontal hydraulic gradients are about 20 ft/mi along Walkers Prairie, but become much greater and range from 80 to 200 ft/mi from near Ford to the southern extent of the Lower aquifer. In Camas Valley, water-level altitudes within the Lower aquifer range from about 2,150 to less than 2,050 ft. The horizontal gradient in Camas Valley is about 100 ft/mi.

The location of the groundwater divide for the Lower aquifer is near the surface-water divide for the basin, near Springdale, Washington (Kahle and others, 2010; fig. 6). Its location was determined by measuring water levels in Lower aquifer wells north and south of the surface-water divide in the

Colville River and Chamokane Creek basins, respectively. The groundwater divide is approximately mid-way between two 1,885 ft water-level altitude contours, one in the Colville River basin and one in the Chamokane Creek basin (fig. 6).

Directions of vertical flow were inferred from water-level altitudes in the Upper outwash aquifer and the Lower aquifer where the units overlie one another (Kahle and others, 2010). Near the confluence of Chamokane and Swamps Creeks, downstream of Icebox Canyon, the difference in water levels was almost 60 ft, with a downward gradient from the Upper outwash aquifer to the underlying Lower aquifer. Conversely, just north of the northeastern tip of the Spokane Reservation (fig. 1), the difference in water levels was about 20 ft, with an upward gradient from the Lower aquifer to the overlying Upper outwash aquifer. On Walkers Prairie, midway between the northeastern tip of the Reservation and USGS streamflow-gaging station 12433200 (fig. 1), the difference in water levels was about 4 ft, again with an upward gradient from the Lower aquifer to the overlying Upper outwash aquifer. Based on available water-level data, vertical flow in the basin generally is downward in the high-altitude areas of the side basins and near Swamp Creek. Vertical flow along Walkers Prairie generally is upward. In the Camas Valley, an upward vertical head gradient exists at the western end of Camas Valley, however, about 1.5 mi downvalley to the east, the difference in water levels was about 90 ft, with a downward gradient from the Valley confining unit to the Lower aquifer (Kahle and others, 2010). Flowing wells, downvalley from Ford, indicate upward head gradients from the Lower aquifer to overlying units.

Paired hydrographs for closely spaced wells completed in the Upper outwash aquifer and the Lower aquifer indicate a nearly identical timing of the seasonal rise and decline in water levels with similar, but slightly greater magnitude fluctuations in the Upper outwash aquifer (Kahle and others, 2010). The overall similarity of seasonal fluctuations in water levels indicates that these systems may be fairly well connected or respond similarly to seasonal stresses, despite the thick, continuous Valley confining unit.

Groundwater Discharge

Discharge from the Upper outwash aquifer occurs mostly as pumping from wells and at springs and seeps. A line of springs and seeps along an arcuate bluff west of Ford represents the major discharge zone of the Upper outwash aquifer (fig. 1). Outflow from this region supports year-round flow in Chamokane Creek downstream of Ford. Springs discharging from the Upper outwash aquifer along Swamp Creek and between Icebox Canyon and the confluence of Chamokane and Swamp Creeks (fig. 1) support streamflow in those stream reaches. Just downstream, however, surface flow in the stream disappears during much of the year near the northeastern corner of the Reservation. Discharge from the Lower aquifer occurs as pumping from wells and, in areas of upward flow gradients, as discharge to overlying

hydrogeologic units. At the east end of Camas Valley, groundwater flow likely discharges into overlying sediments and Chamokane Creek near the head of Icebox Canyon. Similarly, discharge at the lower end of the basin, south of Ford, may be upward through overlying units and ultimately into Chamokane Creek.

Springs are an important component of the Chamokane surface-water system and sustain much of the flow during the usually dry months of July to November. Groundwater discharges at a series of large springs at the southern end of Walkers Prairie along an east–west oriented bluff beginning just west of the town of Ford. Much of this hillside seeps water in places causing the stream channel to have swampy reaches. Larger outflows occur at the Dawn Mining Company spring, the Washington Department of Fish and Wildlife spring, the Spokane Tribal Woodworks site, and the Galbraith Springs by the Spokane Tribal Fish Hatchery (Kahle and others, 2010, pl. 1). Chamokane Creek is perennial downstream of these spring discharges and remains so, gaining water from the Upper outwash aquifer, for the rest of its course until its confluence with the Spokane River.

Swamp Creek is the only tributary to Chamokane Creek that is perennial; all other tributaries are intermittent. Swamp Creek differs from the other tributaries in that it does not originate in the highlands. Swamp Creek is spring fed and begins in a swampy area just west of the town of Springdale. Thomas Creek also is spring fed and flows southeast out of the hills at the western edge of Walkers Prairie (Kahle and others, 2010, pl. 1), but terminates in a pond on the floodplain that appears to have no surface-water outlet. Tributaries other than Swamp Creek reach Chamokane Creek as surface flow only during the spring snowmelt season and for short periods after significant rainfall events.

Interactions Between Groundwater and Surface Water

Based on surface-water measurements made in the basin from 2007 to 2008 (Kahle and others, 2010), there is considerable interaction between the near-surface hydrogeologic units and the surface water of the basin. Streamflow gains and losses along many reaches vary from month to month and from year to year. Exceptions to this were determined at the stream reaches with flow supported by perennial springs, most notably downstream of Ford where large springs discharge from the Upper outwash aquifer (Kahle and others, 2010). Other spring-supported areas include Swamp Creek and Chamokane Creek between Ice Box Canyon and its confluence with Swamp Creek (Kessler, 2008).

During the high-flow measurements made for this investigation, gains in streamflow occurred throughout the Camas Valley, with the largest high-flow measurement (522 ft^3/s) made at the mouth of Ice Box Canyon where Chamokane Creek exits Camas Valley. From the mouth of Ice Box Canyon to the northern end of Walkers Prairie, large streamflow losses were recorded, indicating that Chamokane

Creek loses flow directly to the Upper outwash aquifer in that reach. Modest gains occurred along Chamokane Creek through Walkers Prairie, apparently due to inputs from tributary streams rather than groundwater discharge to the creek. An overall small loss of flow was measured downstream of Ford to the gaging station downstream of Chamokane Falls, indicating modest recharge to the Upper outwash aquifer in the lower end of the basin.

In contrast, under low-flow conditions, only two reaches along Chamokane Creek show any significant gains in flow and both are due to springs discharging from the Upper outwash aquifer. The first area is between the outlet of Ice Box Canyon and just upstream of the confluence of Chamokane and Swamp Creeks where gains of about 1 ft^3/s were measured in each of the low-flow measurements in 2007 and 2008 (Kahle and others, 2010). The second area is downstream of Ford to near the gaging station at Chamokane Falls, where gains of about 20 ft^3/s were measured during low-flow measurements. The largest loss in flow during the low-flow measurements, slightly more than 2 ft^3/s, was measured between Chamokane Creek just upstream of its confluence with Swamp Creek and the northeastern end of Walkers Prairie.

Additional comparisons of gain/loss patterns for specific reaches between low-flow and high-flow are valuable and have important implications for groundwater/surface-water interactions. For example, a small loss was measured during low flow, but a large gain was measured during high flow, in the reach upstream of the outlet of Ice Box Canyon. The reach between the outlet of Ice Box Canyon and the confluence of Chamokane and Swamp Creeks gained during low flow and lost during high flow. These examples illustrate changes in the direction of groundwater/surface-water fluxes as a function of hydrologic conditions.

Simulation of Groundwater and Surface-Water Resources

Development of a calibrated, coupled groundwater and surface-water flow model allows for an analysis of the movement of water through the hydrogeologic units that constitute the Chamokane Creek basin aquifer system and the potential simulated effects of stresses (and changes in stresses) on the groundwater and surface-water resources. The USGS coupled groundwater and surface-water flow model GSFLOW (Markstrom and others, 2008) was used to investigate the aquifer-stream interactions, provide water budgets, and simulate the effects of current and potential groundwater pumping on Chamokane Creek. The version of GSFLOW (v.1.1.5) used in this study is an integration of the USGS Precipitation-Runoff Modeling System (PRMS; Leavesley and others, 1983; 2005) with the 2012 version of the USGS Newton formulation of the Modular Groundwater Flow Model (MODFLOW-NWT; Niswonger and others, 2011). GSFLOW can be run in PRMS-only, MODFLOW-only, or integrated

mode. Running GSFLOW in PRMS-only and MODFLOW-only modes allows the model to be calibrated and tested in a sequential approach for methodical and efficient model calibration.

The constructed model is a transient model that simulates the period October 1979 through September 2010 [water years 1980–2010; herein, a stated year is a water year (period from October 1 through September 30]. Water year 1980 was selected as a suitable starting point because it was an average year for precipitation and streamflow, groundwater development was minimal, there is a paucity of data before 1980, and the simulation period represents conditions since the 1979 adjudication. The model is suited for providing information on the effects of regional stresses on the groundwater and surface-water flow system during this time period.

Description of Surface-Water Flow Model (PRMS)

PRMS is a physically based, distributed-parameter model designed to simulate precipitation and snowmelt runoff. Major advantages of this system include the ability to (1) simulate the moisture balance of each component of the hydrologic cycle, (2) account for heterogeneous physical characteristics of a basin, and (3) appropriately simulate both mountainous and flat areas.

A basin is conceptualized as an interconnected series of reservoirs whose collective output and interaction produces the total hydrologic response (fig. 7). These reservoirs include interception storage in the vegetation canopy, storage in the soil zone, subsurface storage between the surface of a basin and the water table, and groundwater storage. Lateral subsurface flow (or interflow) is considered to be the relatively rapid movement of water flowing from shallow soils to a stream channel. For non-integrated simulations (that is, PRMS-only simulations), groundwater flow is represented by the PRMS groundwater reservoir. Flow to a groundwater reservoir comes from the overlying soil zone and a subsurface reservoir. The groundwater reservoir is considered the source of all baseflow during PRMS-only simulations. During integrated simulations, the groundwater reservoir is replaced by MODFLOW. The application of the model for this study was run on a daily time step. The system inputs included daily precipitation and daily maximum and minimum air temperature; all other atmospheric forcing variables (for example, short-wave radiation) were estimated on the basis of maximum and minimum air temperatures and other model variables. For PRMS-only simulations, streamflow at a basin outlet is the sum of overland and shallow subsurface flow that occurs within the hydrologic response unit (HRU) reservoirs, and groundwater flow from the PRMS groundwater reservoirs. Lateral flows among HRUs, groundwater reservoirs, and streams are routed using the PRMS cascade-flow procedure, where flow directions are determined from slopes in the land surface derived from a digital elevation model (DEM).

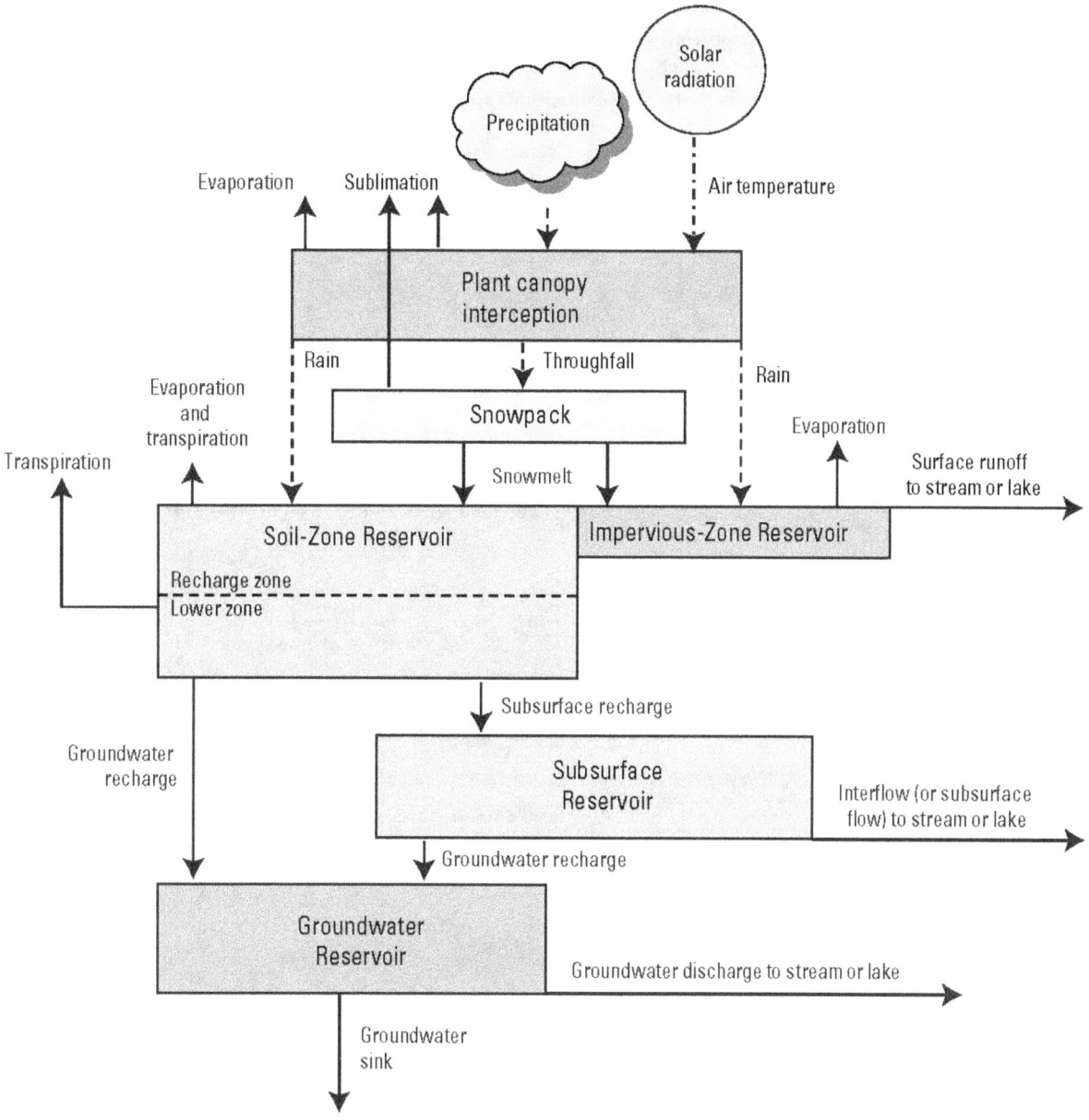

Figure 7. Precipitation-Runoff Modeling System (modified from Leavesley and others, 1983).

Surface runoff and infiltration in the daily time step are computed using a variable source-area approach (Hewlett and Hibbert, 1967; Hewlett and Nutter, 1970). Surface runoff is related to a dynamic source area that expands and contracts in response to rainfall conditions, and to the capability of the soil mantle to store and transmit water (Troendle, 1985). As conditions become wetter, the proportion of precipitation diverted to surface runoff increases, while the portion that infiltrates to the soil zone and subsurface reservoirs decreases. Daily infiltration is computed as the net precipitation minus surface runoff. Precipitation retained on the land surface is modeled as surface-retention storage. Once the maximum retention storage is satisfied, excess water becomes surface runoff. The retention storage is depleted by evaporation in the absence of snow.

Precipitation that falls through the crown canopy infiltrates the soil zone. The soil zone is conceptualized as a two-layer system. Moisture in the upper soil (or recharge) zone and in the lower soil zone is depleted through root uptake and seepage to lower zones. Evaporation also depletes the upper soil zone of moisture. The depths of the soil zones are defined on the basis of water-storage characteristics, depth to bedrock, and the average rooting depth of the dominant vegetation.

Potential evapotranspiration (PET) losses were computed as a function of solar radiation, which in turn, was estimated from daily maximum and minimum temperatures and the number of cloudless days (Jensen and Haise, 1963). When soil moisture is nonlimiting, actual evapotranspiration (AET) equals PET. When soil moisture is limiting, AET is computed from PET to AET relationships for soil types as a function of the ratio of current available water in the soil profile to the maximum available water-holding capacity of the soil profile (Zahner, 1967). During integrated simulations, evapotranspiration (ET) can occur beneath the soil zone in the deeper unsaturated and saturated zones where roots extend below the soil zone. Additionally, during integrated simulations groundwater can flow to the soil zone and increase moisture and ET.

In the Chamokane Creek basin, snow accumulation and subsequent melting produces the vast majority of runoff in the spring. Accurate simulation of this yearly cycle is essential for proper agreement between modeled and measured runoff. The simulation model contains a snow module to simulate the initiation, accumulation, and depletion of a snowpack in each HRU. The snow routine requires computed daily shortwave radiation.

In PRMS-only simulations, soil water in excess of field capacity drains to subsurface and groundwater reservoirs. Soil water in excess of field capacity is first directed to a groundwater reservoir based on a user-specified recharge rate. When daily moisture accumulation exceeds this daily rate, excess soil water is directed to the subsurface reservoir. Excess moisture in the subsurface reservoir either percolates to a groundwater reservoir or flows to a discharge point above the water table. Seepage to the groundwater reservoir is computed first from the soil zone, then as a function of a recharge rate coefficient and the volume of water in the subsurface reservoir.

Markstrom and others (2008) describe the new GSFLOW Soil-Zone module used to simulate flow in the soil-zone and subsurface reservoirs developed to facilitate integration between PRMS and MODFLOW and allow for flow through macropores and Dunnian runoff (Dunne and Black, 1970).

Delineation of Basin Physical Characteristics

The physical attributes of the basin were characterized in a format that could be readily used in the modeling process. Digital data exist that describe the topographic features, soils, land use, and vegetation. These spatial features determined in large part the quantity and movement of water throughout a basin, and subsequent steps in the modeling process built upon this initial characterization.

The initial input to define topographic surfaces was a standard 10-meter USGS 7.5-minute DEM of the Chamokane Creek basin. The 10-meter DEM contains regularly spaced cells, 32.8 ft on center, with altitude reported to the nearest 1 ft at each cell.

The HRUs were delineated as 1,000 × 1,000 ft grid cells to directly coincide with the groundwater-model grid. Land-surface altitudes for each HRU were averaged from the original DEM by resampling the DEM to the size of the HRU grid. Commonly in PRMS modeling, HRU delineation is based on the surface-water drainage network (topography, slope, and aspect) and then further divided by two or more flow planes. Matching the PRMS HRUs directly to the underlying groundwater-model grid established a direct relation between the land-surface hydrologic processes and flow to the subsurface zone as represented by the groundwater-model grid array.

The Cascade Module is used to define connections for routing flow from upslope to downslope HRUs and stream segments, and allows for a complex flow path for surface runoff and shallow subsurface flow within the model domain. The Cascade Module allows surface runoff and interflow to satisfy soil-zone storage of downslope HRUs before being added as streamflow (Markstrom and others, 2008). Flow paths start at the highest upslope HRUs and continue through downslope HRUs until reaching a stream segment. The automated cascade algorithm worked well for most of the study area, but areas of gentle relief combined with a 1,000 × 1,000 ft grid-cell altitude created localized swales, or HRUs with no outflowing cascade links (that is, an HRU that does not intersect a stream and has no neighboring HRUs with lower altitude). Thus, HRUs with no outflow links were filled to allow continuous flow from all HRUs to streams. HRUs were filled by incrementally raising the altitude of these HRUs by a small value until an outflow link was established.

Climate Data

The primary climatic factors that affect the groundwater and surface-water flow systems in the Chamokane Creek basin and that are specified in the model are measured daily precipitation and daily maximum and minimum air-temperature time-series data. The period of climate record used in model simulations was 1980–2010. Not all stations existed for the entire period of record. Periods of missing data from any station simply were not used in model simulations.

Measurements of daily precipitation and air-temperature maximums and minimums to the HRUs were distributed across the Chamokane Creek basin to each HRU using a multiple linear regression (MLR) relation for each dependent climate variable using the independent variables of station latitude, longitude, and altitude (module XYZ, developed by Hay and others, 2000). The XYZ module requires monthly lapse rate parameters for the dependent variables precipitation and maximum and minimum temperatures. This two-dimensional parameter has an MLR coefficient for each independent variable [that is, the x-coordinate (x), y-coordinate (y), or altitude (z)] by month, and therefore has an array dimension of 3 × 12.

Daily precipitation totals and minimum and maximum air-temperature data used in the GSFLOW model simulations were collected at climate stations located in the Chamokane Creek basin and surrounding areas. Climate stations operated by the U.S. National Weather Service (NWS) and the Bureau of Reclamation (Reclamation) provided data from a total of four stations (fig. 8, table 1) with varying periods of record.

Only Reclamation's Agrimet station was located within the study area boundary and data from that station only exist for the last 3 years (2008–10) of the model simulation period. Therefore, no meaningful comparison between distributed and measured precipitation and temperature values can be made. Simulated mean monthly precipitation and maximum and minimum air temperatures were compared to estimates from the Parameter-estimation Regressions on Independent Slopes Model (PRISM; Daly and others, 1994, 1997) and adjustments were made to model lapse rates until a reasonable match was achieved.

PRMS Model Parameterization

Mathematically, parameters are defined as numerical constants that are used to simulate variables such as streamflow. These variables are computed by equations during the simulation. PRMS has both distributed and non-distributed parameters. Distributed parameters are attributed to each HRU and describe (1) physiographic characteristics, such as area, slope, and aspect; (2) hydrologic processes within the HRU, such as subsurface or groundwater flow; and (3) climatic input to the HRU, such as precipitation and temperature adjustments. Non-distributed parameters are parameters held constant throughout the basin, such as the Julian date to force snowpack depletion or the temperature that determines the form of precipitation. All PRMS parameters are defined and discussed in depth by Leavesley and others (1983, 1996).

Parameters for the discrete spatial features of the study area were generated using the Geographic Information System (GIS) Weasel toolbox (Viger and Leavesley, 2007). In addition to altitude, slope, and aspect, ancillary information concerning soils, land use and land cover, and vegetation were incorporated to assign further characteristics to each HRU. Digital soil data were obtained from a modified version of general soil maps from the State Soil Geographic Database (STATSGO; U.S. Department of Agriculture, 1994). Parameters from the contiguous U.S. Forest Type Groups map and U.S. Forest Density map provided vegetation information (Zhu and Evans, 1992; Powell and others, 1998). The 40-meter 2001 National Land Cover Data (NLCD) database (Multi-Resolution Land Characteristics Consortium, 2008) was used to determine the dominant vegetation type, percentage of impervious surface, and vegetation canopy density for each HRU.

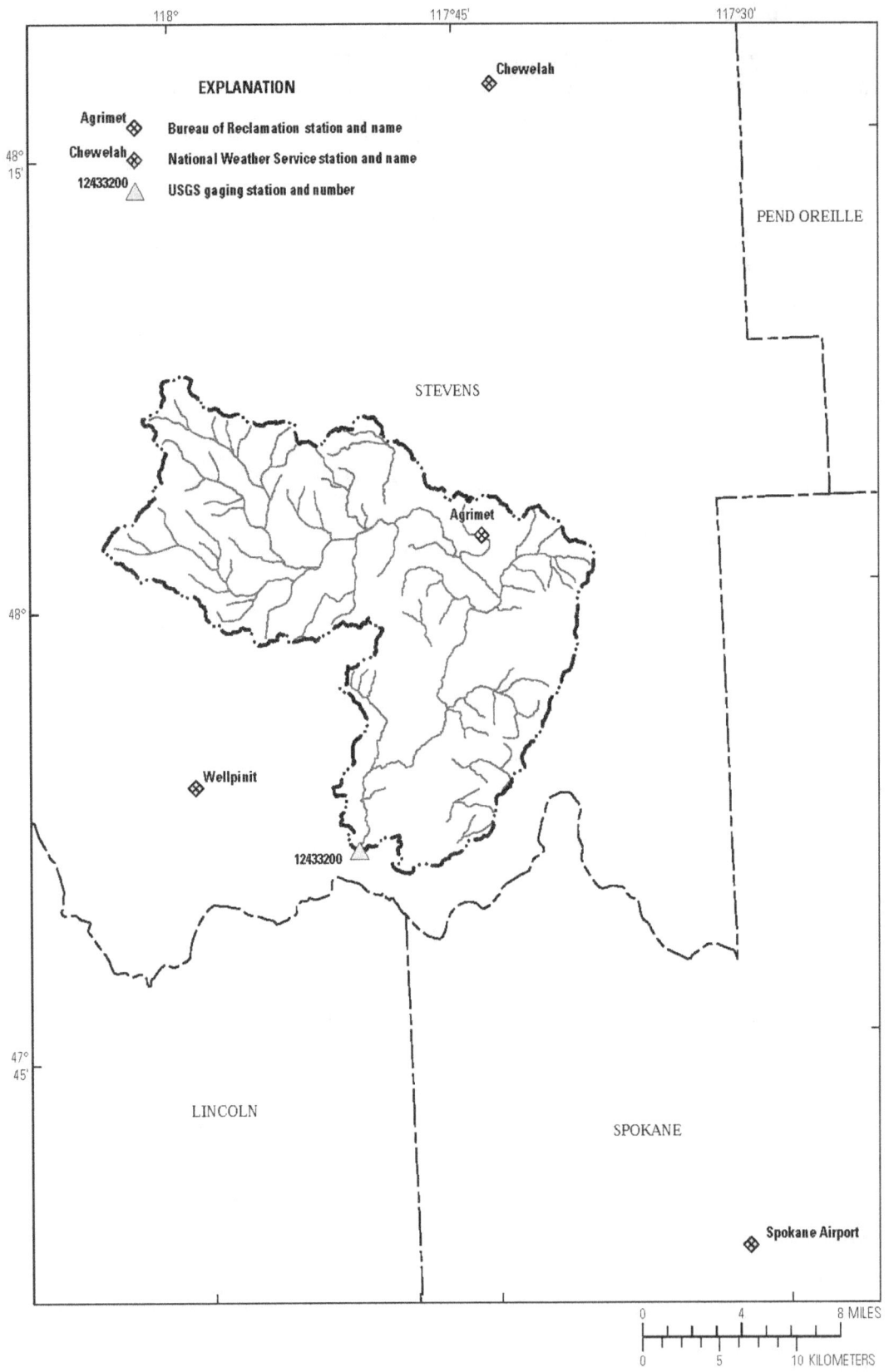

Figure 8.   Climate and stream-gaging data-collection network used for the coupled groundwater and surface-water flow model, Chamokane Creek basin, Washington.

Table 1.   Climate stations used in model simulations, Chamokane Creek basin, Washington.

[**Agency:** NWS, National Weather Service; Reclamation, Bureau of Reclamation. **Altitude:** NAVD 1988, North American Vertical Datum of 1988]

Station name	Agency	Latitude	Longitude	Altitude (feet above NAVD 1988)	Period of record
Spokane Airport	NWS	47°38'00"	117°32'00"	2,360	January 1948—September 2010
Wellpinit	NWS	47°54'00"	118°00'00"	2,490	June 1948—December 2007
Agrimet	Reclamation	48°01'53"	117°44'21"	1,950	November 2007—September 2010
Chewelah	NWS	48°17'00"	117°43'00"	1,670	June 1948—September 2010

Description of Groundwater Flow Model (MODFLOW-NWT)

The Chamokane Creek model used a version of GSFLOW that incorporates MODFLOW-NWT (Niswonger and others, 2011), a standalone program that is intended for solving problems involving drying and rewetting nonlinearities of the unconfined groundwater-flow equation. Application of the Newton method required changes to the internal structure of MODFLOW-2005 (Harbaugh, 2005). MODFLOW-NWT was used to simulate groundwater flow in the basin-fill deposits and basalt and bedrock units of the aquifer system, and the interaction of the groundwater-flow system with surface-water features. MODFLOW–NWT uses datasets describing the hydrogeologic units, unsaturated zone, boundary conditions, water use, initial conditions, and hydraulic properties, and calculates hydraulic heads at discrete points (nodes in a model cell) and flows within the model domain. Similar to PRMS-only simulations, MODFLOW-NWT can be run independent of PRMS (referred to as MODFLOW-only simulations).

Upstream Weighting Package

The Upstream Weighting (UPW) Package is an internal flow package for MODFLOW-2005 intended to be used with the Newton Solver (NWT) for problems involving drying and rewetting nonlinearities of the unconfined groundwater-flow equation. The UPW Package treats nonlinearities caused by the drying and rewetting of model cells by use of a continuous function rather than the discrete function approach to drying and rewetting that is used in the Block-Centered Flow (BCF), Layer Property Flow (LPF), and Hydrogeologic-Unit Flow (HUF) Packages (Anderman and Hill, 2000; Harbaugh, 2005). This further enables application of the Newton solution method for unconfined groundwater-flow problems because conductance derivatives required by the Newton method are smooth over the full range of head for a model cell. A complete description of the UPW Package can be found in Niswonger and others (2011).

Newton Solver

The Newton method is a commonly used numerical method in the earth sciences to solve nonlinear differential equations. The method is advantageous because many of the recently developed MODFLOW packages apply nonlinear boundary conditions. Additionally, the Newton method has shown to be particularly beneficial in solving problems representing unconfined aquifers. The NWT Solver includes two previously developed asymmetric linear solver options—a generalized-minimum-residual (GMRES) Solver and an Orthomin / stabilized conjugate-gradient (CGSTAB, called χMD) Solver. The Chamokane Creek GSFLOW model used the χMD matrix solver. It was found during the course of this study that the Newton method provided greater model stability and improved model convergence compared to the solvers used in the standard MODFLOW-2005 code. A complete description of the Newton formulation and χMD Solver can be found in Niswonger and others (2011).

Unsaturated-Zone Flow

The Unsaturated-Zone Flow (UZF1; Niswonger and others, 2006) Package for MODFLOW-2005 and MODFLOW-NWT simulates water flow and storage in the unsaturated zone and partitions infiltration across the land surface into evapotranspiration and recharge. The package also accounts for land-surface runoff to streams and lakes. A kinematic wave equation for unsaturated flow is solved by the method of characteristics to simulate vertical unsaturated flow. The approach assumes that unsaturated flow occurs in response to gravity potential gradients only and ignores negative potential gradients; the approach further assumes uniform hydraulic properties in the unsaturated zone for each vertical column of model cells between the base of the soil zone and water table. The Brooks-Corey function is used to define the relation between unsaturated hydraulic conductivity and water content. Variables used by the UZF1 Package include initial and saturated water contents, saturated vertical hydraulic conductivity, and an exponent in the Brooks-Corey function.

Residual water content is calculated internally by the UZF1 Package on the basis of the difference between saturated water content and specific yield.

The UZF1 Package is a substitution for the Recharge and Evapotranspiration Packages of MODFLOW-2005. The UZF1 Package differs from the Recharge Package in that an infiltration rate is applied at land surface instead of a specified recharge rate directly to groundwater. The applied infiltration rate is further limited by the saturated vertical hydraulic conductivity. The UZF1 Package differs from the Evapotranspiration Package in that evapotranspiration losses are first removed from the unsaturated zone above the evapotranspiration extinction depth, and if the demand is not met, water can be removed directly from groundwater whenever the depth to groundwater is less than the extinction depth. The UZF1 Package also differs from the Evapotranspiration Package in that water is discharged directly to the soil zone for integrated simulations whenever the altitude of the water table exceeds land surface. Water that is discharged to land surface flows to the PRMS soil zone for integrated simulations, and for MODFLOW-only simulations, applied infiltration in excess of the saturated vertical hydraulic conductivity, may be routed directly as inflow to specified streams; otherwise, this water is removed from the model.

Recharge and discharge were assigned to the highest active model layer in each vertical column of the grid. Groundwater discharge to land surface in the area of Swamp Creek was routed to the nearby streamflow-routing segment representing Chamokane Creek during the MODFLOW-only simulation; and flowed to the soil zone during integrated simulations. The Brooks-Corey epsilon coefficient, which is used to define the relation between unsaturated hydraulic conductivity and water content, and the saturated water content of the unsaturated zone were set to constant values of 3.5 (dimensionless) and 0.18 ft^3/ft^3, respectively. The initial water content for each vertical column of cells ranged from 4.5×10^{-4} to 1.4×10^{-3} ft^3/ft^3.

Spatial and Temporal Discretization

The MODFLOW program uses datasets that describe the hydrogeologic units, recharge, discharge, and conceptual model of the groundwater-flow system, and calculates hydraulic heads at discrete points (nodes) and flow within the model domain. The program requires that the groundwater-flow system be subdivided, vertically and horizontally, into rectilinear blocks called cells. The hydraulic properties of the material in each cell are assumed to be homogeneous. The Chamokane Creek basin study area was subdivided into a horizontal grid of 106 columns and 102 rows; cells are a uniform 1,000 ft per side (fig. 9). The cell size and uniform grid spacing were chosen to reflect the regional scale of this study.

Vertically, the model domain was subdivided into eight model layers. Six model layers were used to simulate the variably saturated unconsolidated sediments that overlie

the bedrock and two layers were used to simulate the basalt and bedrock units. The extents of active cells in each layer correspond to the presence of the hydrogeologic unit(s) simulated in that layer as shown in figures 10A–10E. For discontinuous units within the active model domain, the model layer was assigned a thickness of 1 ft and given the hydraulic properties representative of the material that was present. Note that the size of the grid cells does not imply precision at that scale.

Hydrogeologic unit	Model layer
Upper outwash aquifer	1 and 2
Landslide unit	3
Valley confining unit	4 and 5
Lower aquifer	6
Basalt and bedrock unit	7
Bedrock unit	8

The Upper outwash aquifer is present throughout much of the length of the valley and therefore plays an important role in groundwater/surface-water interaction with Chamokane Creek. MODFLOW represents the exchange of water between the stream and the groundwater system as a function of stream geometry and the difference between the head in the stream and the head at the center of an adjacent underlying model cell. To reduce errors produced by this representation, the Upper outwash aquifer unit was subdivided into two model layers (layers 1 and 2). Layer 1, where present, is a uniform 20 ft thick and layer 2 is the remainder of the mapped thickness, ranging from 1 to 350 ft. The Valley confining unit also is present throughout much of the valley and reaches a thickness of 440 ft. To better simulate the vertical gradients throughout the valley, the Valley confining unit was divided into two layers of equal thickness.

Bedrock has low permeability except for where it is fractured, but the fractures are at too small a scale to be represented in a regional model, and little is known about the hydraulic properties of the bedrock at depth. Owing to these uncertainties and necessary simplifications, layer 7 was assigned a constant thickness of 200 ft, roughly the average thickness of the basalt, where present. The bottom of layer 8 was set at a constant 800 ft above sea level, resulting in layer 8 thickness ranging from 226 to 3,466 ft.

The combination of lateral and vertical discretization resulted in 86,496 cells within the model grid, of which 19,656 cells are active. The active cells include an area of 180 mi^2 and comprise 56 mi^3 of aquifer-system material, including the basalt and bedrock units. Total model domain thickness ranged from 656 to 3,666 ft, with an average thickness of 1,641 ft. All model layers were simulated as convertible (variable transmissivity based on head in the cell) except for layer 8 in which transmissivity was held constant in time.

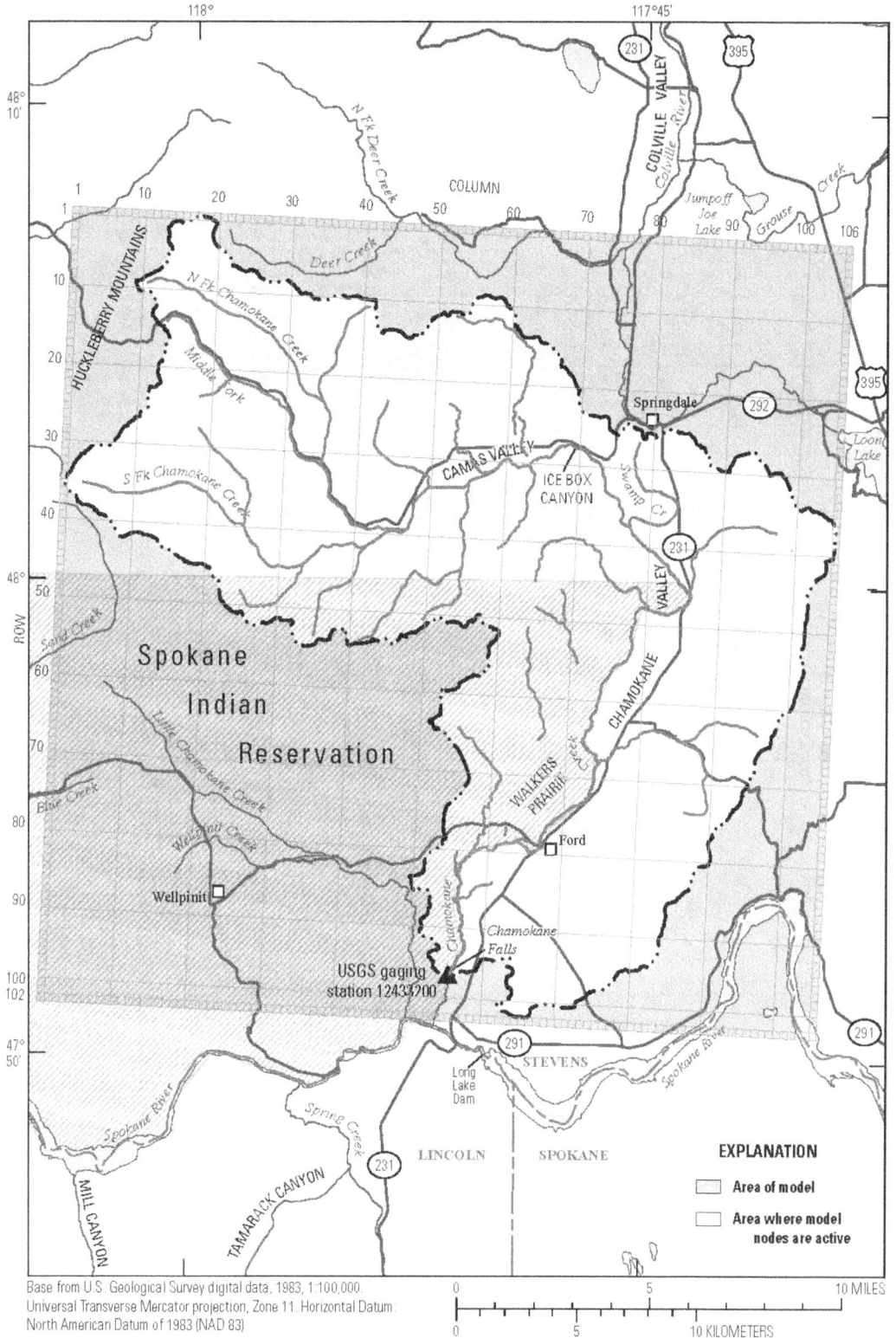

Figure 9. Location and extent of the coupled groundwater and surface-water flow model, Chamokane Creek basin, Washington.

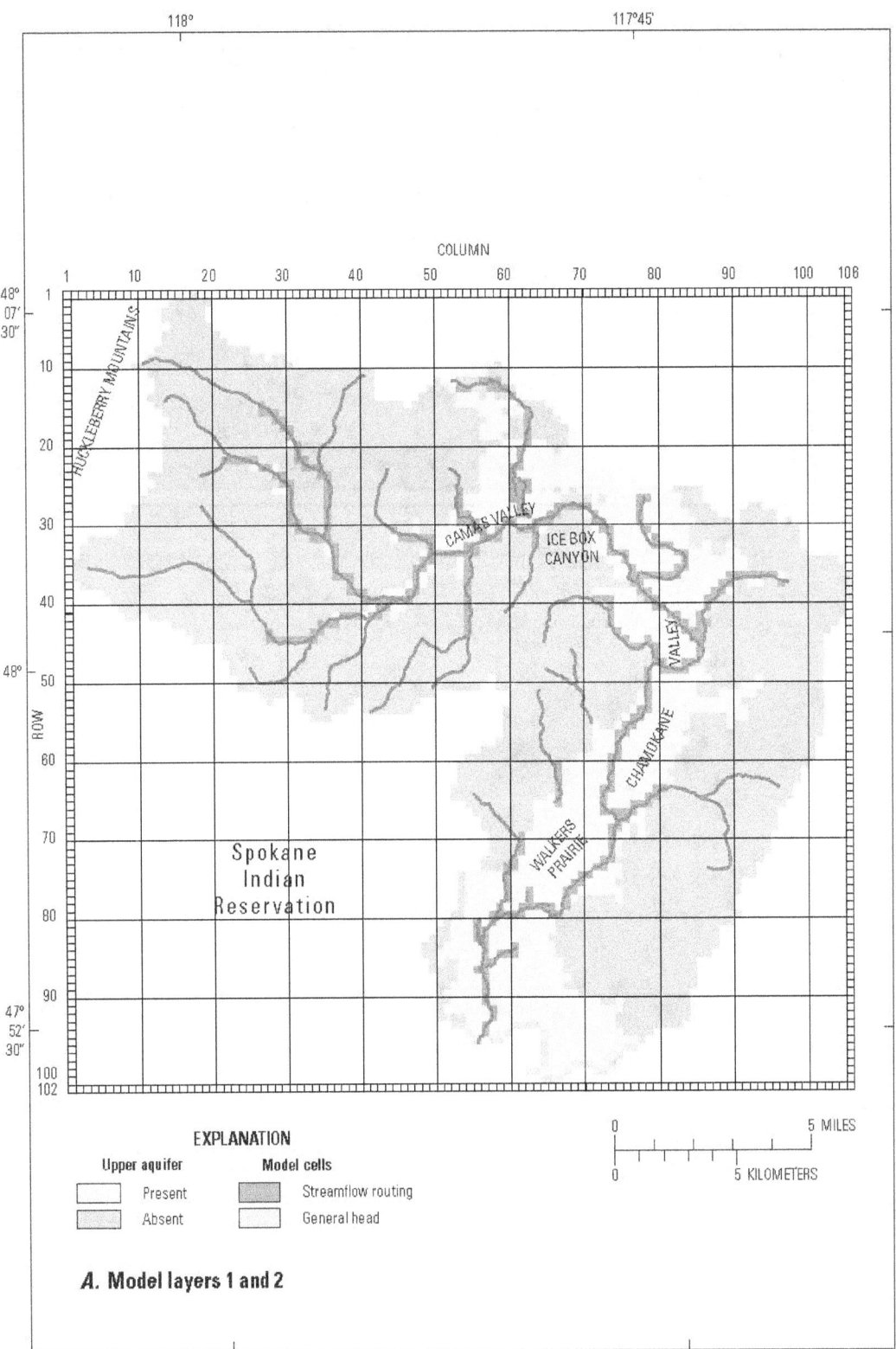

Figure 10.   Areal extents of model layers and locations of streamflow routing and general head cells, Chamokane Creek basin, Washington.

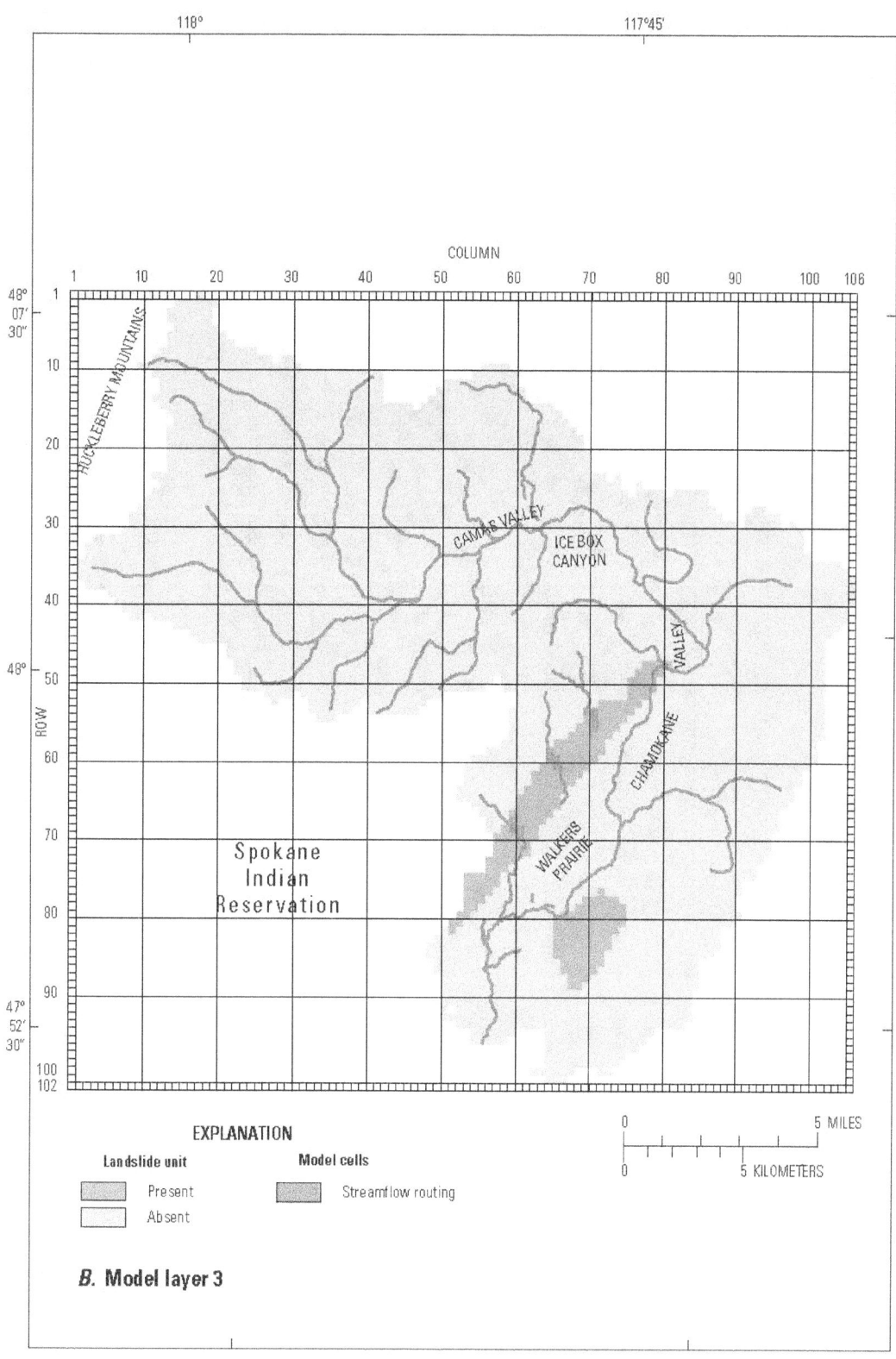

EXPLANATION

Landslide unit

Present

Absent

Model cells

Streamflow routing

0   5 MILES

0   5 KILOMETERS

B. Model layer 3

Figure 10.—Continued

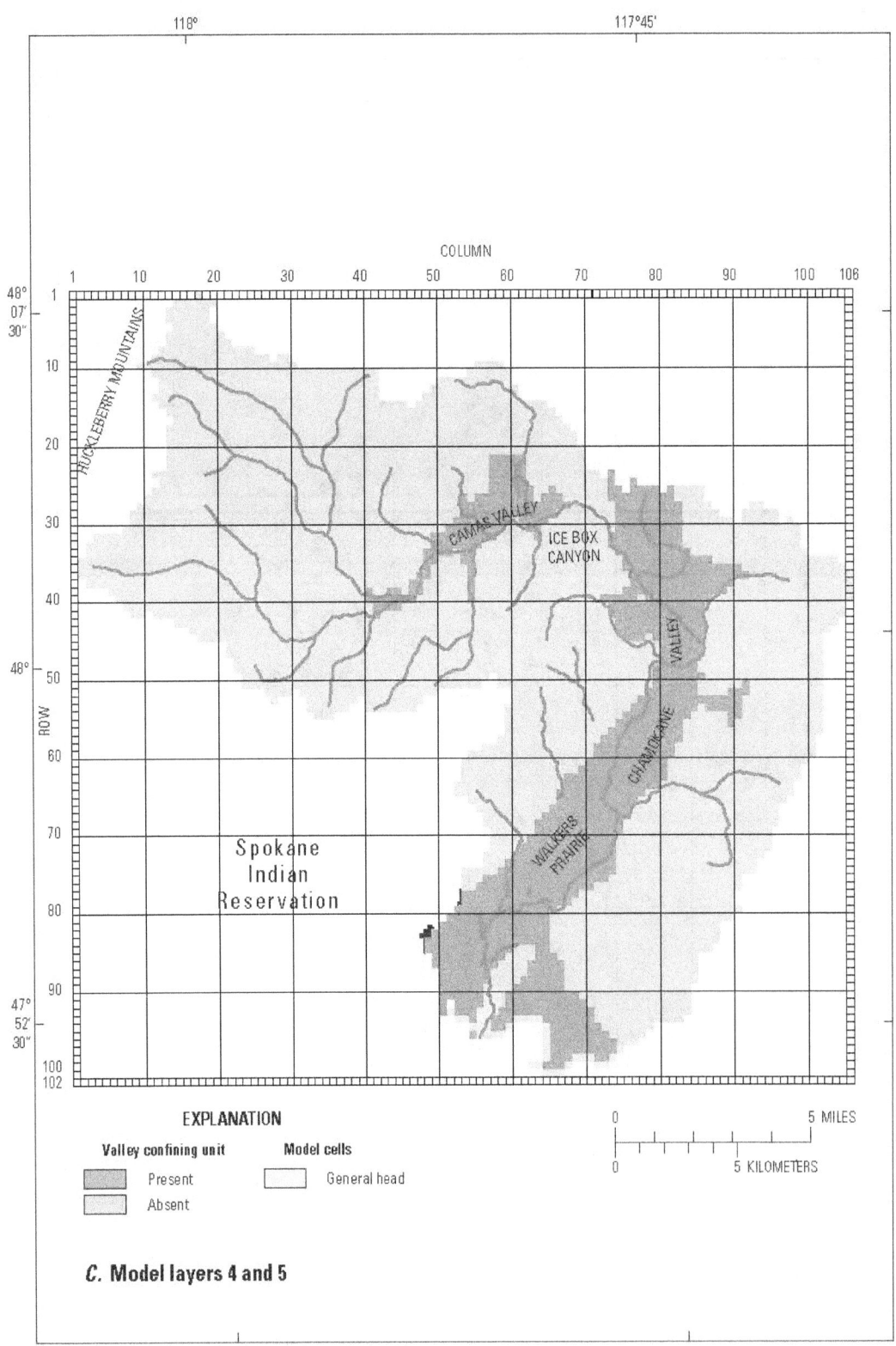

C. **Model layers 4 and 5**

Figure 10.—Continued

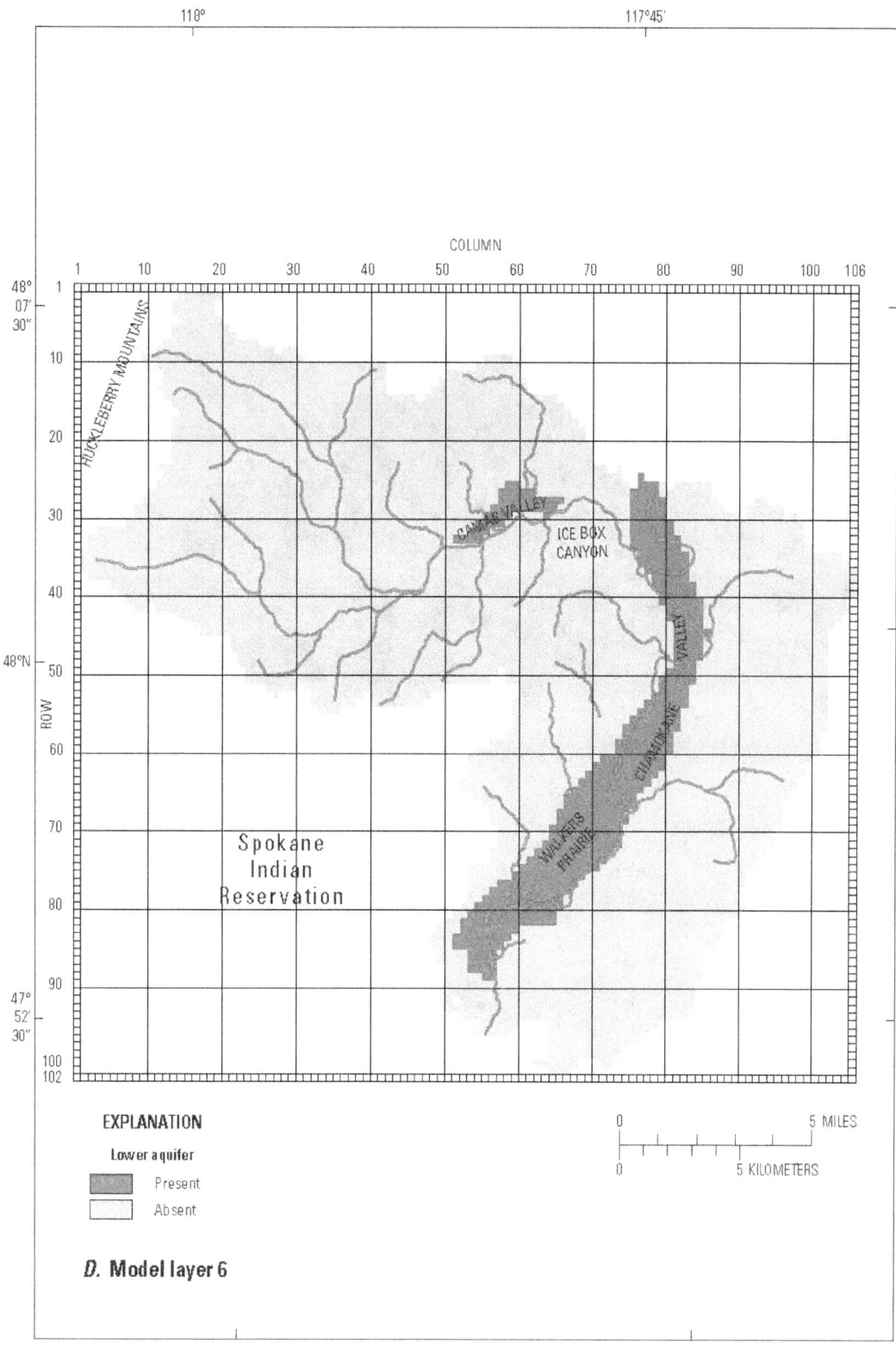

EXPLANATION

Lower aquifer

Present

Absent

D. Model layer 6

Figure 10.—Continued

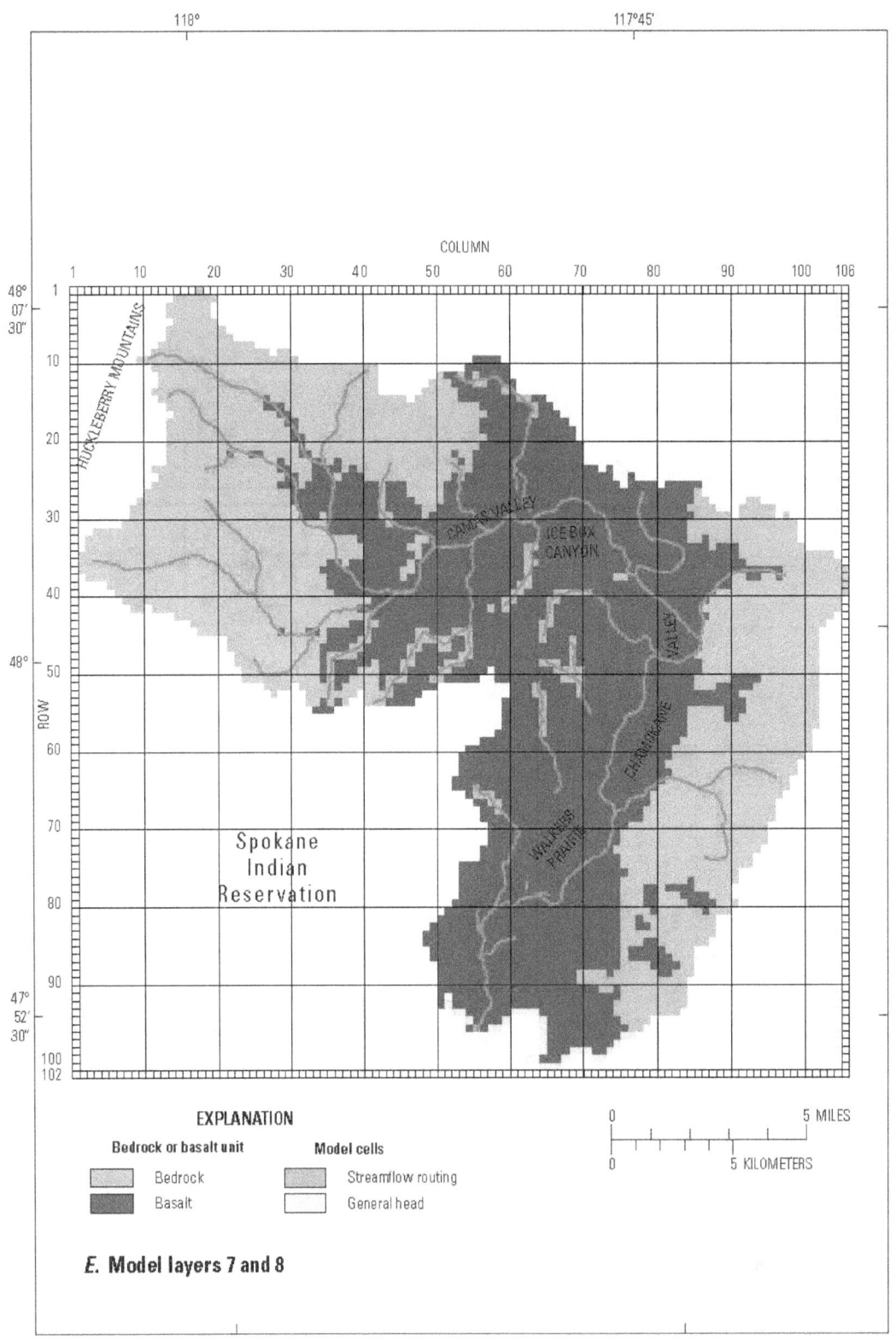

EXPLANATION

Bedrock or basalt unit

Bedrock

Basalt

Model cells

Streamflow routing

General head

0   5 MILES

0   5 KILOMETERS

E. **Model layers 7 and 8**

Figure 10.—Continued

Time discretization in GSFLOW has two levels of division. The finer discretization is called a time step, and a time step is constant at 1 day. The coarser time discretization is called a stress period and it includes 1 or more time steps. Precipitation and temperature vary on a time step (daily) basis, and accordingly, heads and flows can change during each time step. Pumping rates change on a stress period basis. Daily time steps are sufficiently small for MODFLOW to converge to a solution for most simulations (Markstrom and others, 2008). The simulation period extends from October 1, 1979, to September 30, 2010, for a total of 31 water years (1980–2010) and 62 transient stress periods, within which groundwater pumping and surface-water diversions and returns are constant. The length of each stress period was 6 months (October 1–March 31 and April 1–September 30). The stress periods closely matched normal irrigation, stockwatering, and domestic water uses to allow variations in groundwater pumping that matched the seasonal variability. The long simulation period allows for a temporal assessment that accounts for some changes in pumping and a large range in climatic conditions and, thus, large ranges in simulated streamflow and groundwater recharge.

The first stress period in the MODFLOW Discretization file is steady state and no computations pertaining to PRMS are executed for the initial stress period (MODFLOW-only). All model fluxes, including gravity drainage beneath the soil zone, were specified for the steady-state simulation. Results from the first steady-state stress period were used as the initial conditions for the subsequent transient stress periods.

Hydrogeologic Framework

The three-dimensional digital hydrogeologic framework developed for the groundwater-flow model is based on the primary data used by Kahle and others (2010): DEM, geologic maps, cross sections, and lithologic well logs. These data types were manipulated with stratigraphic analysis software and a GIS.

The electronic data were assembled into a single three-dimensional, spatially distributed hydrogeologic representation using GIS for incorporation into the groundwater-flow model. Existing data included (1) surficial geology maps, (2) surficial extents of the four unconsolidated hydrogeologic units, (3) well-log point values and thickness contours for tops and thicknesses of the four unconsolidated units, (4) well-log point values for the top of bedrock and basalt, and (5) nine hydrogeologic sections.

The modeled eight-layer, three-dimensional grid was then compared to the hydrogeologic sections and adjusted where appropriate. An effort was made to honor the geologist's interpretation so the model construction was as representative as possible. Large data gaps and the regional scale of the groundwater model created some discrepancies, but the method described above created a reproducible hydrogeologic representation that was used to create the hydrogeologic framework for the model.

Boundary Conditions

Boundary conditions define the locations and manner in which water enters and exits the active model domain. The conceptual model for the aquifer system is that water enters the system as recharge from precipitation (rainfall and snowmelt) and exits the system as streamflow, evapotranspiration, and groundwater pumpage. The boundaries of the model coincide as much as possible with natural hydrologic boundaries. Three types of model boundaries were used: no-flow boundaries (groundwater divides), head-dependent flux boundaries (streams and general-head boundaries), and specified-flux boundaries (pumpage, stream inflows, and surface-water diversions).

No-Flow Boundaries

The model boundaries that coincide with the natural landscape divides are simulated as no-flow boundaries because they are assumed to be groundwater divides. Water-level data, where available, also suggest the presence of groundwater divides (Kahle and others, 2010) at these locations. Major topographic divides coincide with the northern, eastern, and western model boundaries. The topographic divides are exposed bedrock. These divides are the crest of the Huckleberry Mountains to the northwest, a low-altitude drainage divide with the north flowing Colville River to the north, and drainage divides with Little Chamokane Creek and Little Spokane River to the west and east, respectively.

Head-Dependent Flux Boundaries

Streams

The exchange of groundwater and surface water is an important hydrologic process in the Chamokane Creek basin flow system and, to the extent possible, the model was constructed to capture this process. Chamokane Creek and its tributaries were simulated using the MODFLOW Streamflow-Routing (SFR2) Package to route streamflow and calculate river-aquifer exchanges, and includes the ability to simulate unsaturated flow beneath intermittent and ephemeral streams (Niswonger and Prudic, 2005). The model has 58 simulated stream segments and 779 simulated reaches that are coincident with the underlying MODFLOW cells; the locations of the simulated stream cells are shown in figures 10A, 10B, and 10E.

The exchange of water between streams and groundwater is controlled by the difference in the groundwater level and stream stage in each cell and by the hydraulic properties of the streambed at the river-aquifer boundary in each cell, which is represented in the model by a user-specified streambed conductance term. The depth of each stream within each reach was computed using Manning's equation for open channel flow assuming an eight-point cross section. Average stream depth and stream width for the cross sections were based on mean annual streamflow from the USGS National Hydrography Dataset (http://nhd.usgs.gov/index.html) and regression equations determined by Magirl and Olsen (2009).

The simulated quantity of water moving between the groundwater and surface-water systems is equal to the product of streambed conductance and the simulated head difference between the stream and underlying model hydrogeologic units. Initial values of streambed conductance were based on stream length (determined using GIS) and width (Magirl and Olsen, 2009), estimated streambed hydraulic conductivity, and streambed thickness. Initial estimates of streambed hydraulic conductivity were based on Conlon and others (2003) and adjusted during model calibration. Streambed thickness was set to 1 ft for all stream reaches. The model internally multiplies the hydraulic conductivity value (ft/d) by the stream reach length (ft) and wetted width (ft), divided by the streambed thickness (ft), resulting in the streambed conductance (ft²/d). For routing streamflow, a constant value of 0.04 was used for Manning's coefficient.

Streambed altitudes for Chamokane Creek and its tributaries were determined using the average land-surface altitude of the upstream and downstream location of the stream at each model cell boundary (from the USGS 10-meter DEM) minus stream depths estimated by Magirl and Olsen (2009). Some inaccuracy was introduced in the simulation of groundwater flow to and from the streams by using average stream stages and simulating average groundwater altitudes within model cells. This uncertainty was not deemed a problem in areas of gentle relief, but uncertainty was introduced in areas of steep terrain and incised canyons (with seepage faces contributing to streamflow). Issues related to model grid size are discussed in the section, "Model Uncertainty and Limitations."

General-Head Boundaries

The location of the USGS streamflow-gaging station, Chamokane Creek below Falls, near Long Lake (12433200; fig. 1), is considered the basin outlet in this study. Nearly all of the total flow measured at the gage during the low-flow time of year is provided by groundwater discharge, mostly from large springs and numerous seeps in the section downstream of Ford and from outflow from hatcheries, which is a combination of water from wells and springs. Chamokane Falls, where Chamokane Creek flows over a bedrock outcrop, is about 1.5 mi upstream of the confluence with the Spokane River, and the model boundary at this location is not a strict groundwater divide, because the Upper outwash aquifer, Valley confining unit, basalt, and lower bedrock units are present below the Chamokane Falls. These units, with the exception of the lower bedrock, are largely unsaturated and groundwater flow leaving the study area is considered negligible. To test this conceptual model, the MODFLOW general-head boundary (GHB) package was used to simulate groundwater outflow along the southern extent of the model. The model includes 93 general-head boundary cells that are assigned to model layers 2, 5, 7, and 8 (figs. 10A, 10C, and 10E).

Flow into or out of a GHB cell within the groundwater system is proportional to the product of the specified GHB conductance term and the difference between the simulated hydraulic head in the cell and the specified stage of the GHB cell. The specified stages were set at 1 ft above the cell bottoms for the southern boundary. All stages were held constant throughout the simulation period. The GHB conductance is a function of the surrounding hydrogeologic material and the area. The conductance was initially set at 618.0 (layer 2), 8.0 (layer 5), and 1 ft²/d (layers 7 and 8).

Specified-Flux Boundaries

Two types of specified fluxes were simulated in the model: (1) stream inflows (surface-water return flows) and outflows (surface-water diversions) and (2) groundwater withdrawals (pumpage). If simulated streamflow in a particular reach is insufficient to supply the diversion rate then the diversion is reduced to the available amount of flow. Similarly, if the specified pumping rates draw the water table down to the base of the aquifer then the pumping rate is automatically reduced to an amount that can be supplied by the aquifer.

Surface-Water Return Flows and Diversions

Seasonal (October–March or April–September) inflows (return flows) and outflows (diversions) to and from simulated stream segments were estimated from measured streamflow and locally reported water use. The Chamokane Creek model includes seasonally aggregated streamflow diversions (55) and returns (2). Diversions represent surface-water uses for water-right claims (4) and stock watering (51).

A groundwater-right claim is a document declaring a claim for water use, and may be valid if it describes a groundwater use started before July 1945. Most claims were filed during an open filing period authorized by the 1967 Water Right Claims Registration Act or the 1998 reopening of the Act. For a non-exempt groundwater use that postdates 1945, a right can only be granted through judicial processes involving filing applications, obtaining permits, and field inspections of the well in question. A favorable ruling results in the granting of a water right. In 2010, 19 claims (pre-1945 uses) were in effect in the Chamokane Creek basin (John Covert, Washington State Department of Ecology, written commun., March 2011). Water-right claims diversions were active for all stress periods; each claim was assigned a constant rate of 350 gal/d, which is an average household use (Washington Administrative Code 173-503-073). Claims that were not located near a simulated stream (for example, springs) were simulated as a shallow well.

Diversions for stock watering were estimated from the number and type of livestock, period of watering (for example, 2001–present), and seasonality of watering (year round, October–March, or April-September; Brian Crossley, Spokane Tribe Natural Resources, written commun., March 2011).

Rates of diversions were estimated from the Washington State Department of Ecology (WaDOE) Water Rights Processing manual (John Covert, Washington State Department of Ecology, written commun., March 2011).

Return flows to the streams consist of outflows from the Tribal and State hatcheries, and are equal to the groundwater-pumping rates at each hatchery. The combined total of water pumped for the two hatcheries for aquaculture for 2007 was about 1.15 billion gal. Groundwater pumpage by the hatcheries increased by about a factor of 5 from 1997 to 2007 in response to increased water use at the Tribal hatchery beginning in 2003 and drilling of a production well and subsequent use of groundwater at the Washington Department of Fish and Wildlife hatchery beginning in 2005 (Kahle and others, 2010).

Groundwater Pumping

Groundwater pumping in the Chamokane Creek basin was estimated during the first part of this study for 1980–2009 (Kahle and others, 2010). The major uses of water in the Chamokane Creek basin are for domestic purposes (single-family drinking and household use, lawns, and small gardens), public water supply, fish-hatchery operations, irrigation, and commercial and industrial uses. Some surface water may be withdrawn from Chamokane Creek for these purposes, but the primary source of water was considered groundwater.

Additional water-use estimates became available during Phase 2 of the study and were added to the previous estimates for simulation in the flow model. The additional categories of simulated groundwater pumping are surface-water claims that were simulated as a shallow groundwater withdrawal and stock watering (see section, "Surface-Water Return Flows and Diversions"). Groundwater-use estimates from Phase 1 of this study (Kahle and others, 2010) were refined or recategorized. For example, the commercial/industrial use at the Dawn Mining Company was mostly diverted springflow, instead of groundwater withdrawals, and some of the agricultural irrigation was assigned to claims or permit-exempt wells. Although rates for individual water-use categories changed, total annual groundwater pumpage remained similar. Locations and rates of simulated net groundwater pumpage, by category, are shown in figure 11 and table 2. The rates are net pumpage because they account for both groundwater pumping and returns from septic recharge and non-consumptive outdoor water use. Claims registry rates for the October to March stress periods are negative because they include septic returns from claims for surface-water diversions, which are greater than the claims for groundwater withdrawals.

In 1980, total net simulated annual pumpage in the basin was about 662 acre-ft (0.9 ft^3/s). By 2010, total net annual pumpage was estimated to be about 4,047 acre-ft (5.6 ft^3/s).

Hatchery operations account for about 80 percent of 2010 pumpage. The Tribal fish hatchery accounted for about 66 percent of 2010 pumpage and the Washington Department of Fish and Wildlife hatchery accounted for about 14 percent. Both hatcheries withdraw water from the Lower aquifer. Public water supply accounts for about 18 percent. Irrigation is a larger source of groundwater pumpage during the April through September stress periods than it is during the October through March stress periods.

Locations of permit-exempt pumpage shown in figure 11 are estimated from the WaDOE online well log database (Washington State Department of Ecology, 2003). All existing well logs are located to the nearest Township, Range and quarter section or quarter of a quarter section. The well completion date was used to assign the well to the appropriate stress period. Wells in existence prior to 1980 were used throughout the model simulation period. Not all wells in the WaDOE database are actively being used and not all active wells are in the WaDOE database, so the locations and number of permit-exempt wells are estimates, although they represent the general timing and spatial distribution of groundwater development.

Rates of permit-exempt pumpage presented in table 2 are net withdrawals from the groundwater system and are based on pumping estimates from Kahle and others (2010). Septic recharge was estimated based on a modified method outlined in Vaccaro and Olsen (2007) and Ely and others (2011). For this study, 90 percent of pumpage for indoor use was returned as septic recharge; thus about 52 percent of the annual pumpage becomes recharge, resulting in an additional mean annual septic return value of about 0.04 ft^3/s (26 acre-ft). The septic return was input into the model as an injection well located in the uppermost active model layer at the cell location (row-column) of each exempt well or claims registry use.

Initial Hydraulic Properties

The initial hydraulic properties of horizontal hydraulic conductivity (K_h), vertical hydraulic conductivity (K_v), anisotropy (K_h:K_v), and specific storage were assigned on the basis of values tabulated from previous studies (Kahle and others, 2003, 2010) and analysis of specific-capacity data (Kahle and others, 2010). Uniform values were used to simplify model construction and calibration, and to regionalize hydraulic properties where appropriate.

Hydraulic properties were assigned and initial model simulations were evaluated based on simulated heads and streamflow, numerical closure, and model budget error. Minor adjustments were then made to parameter values to improve model fit. These adjusted values are considered the initial hydraulic properties for calibrating the model.

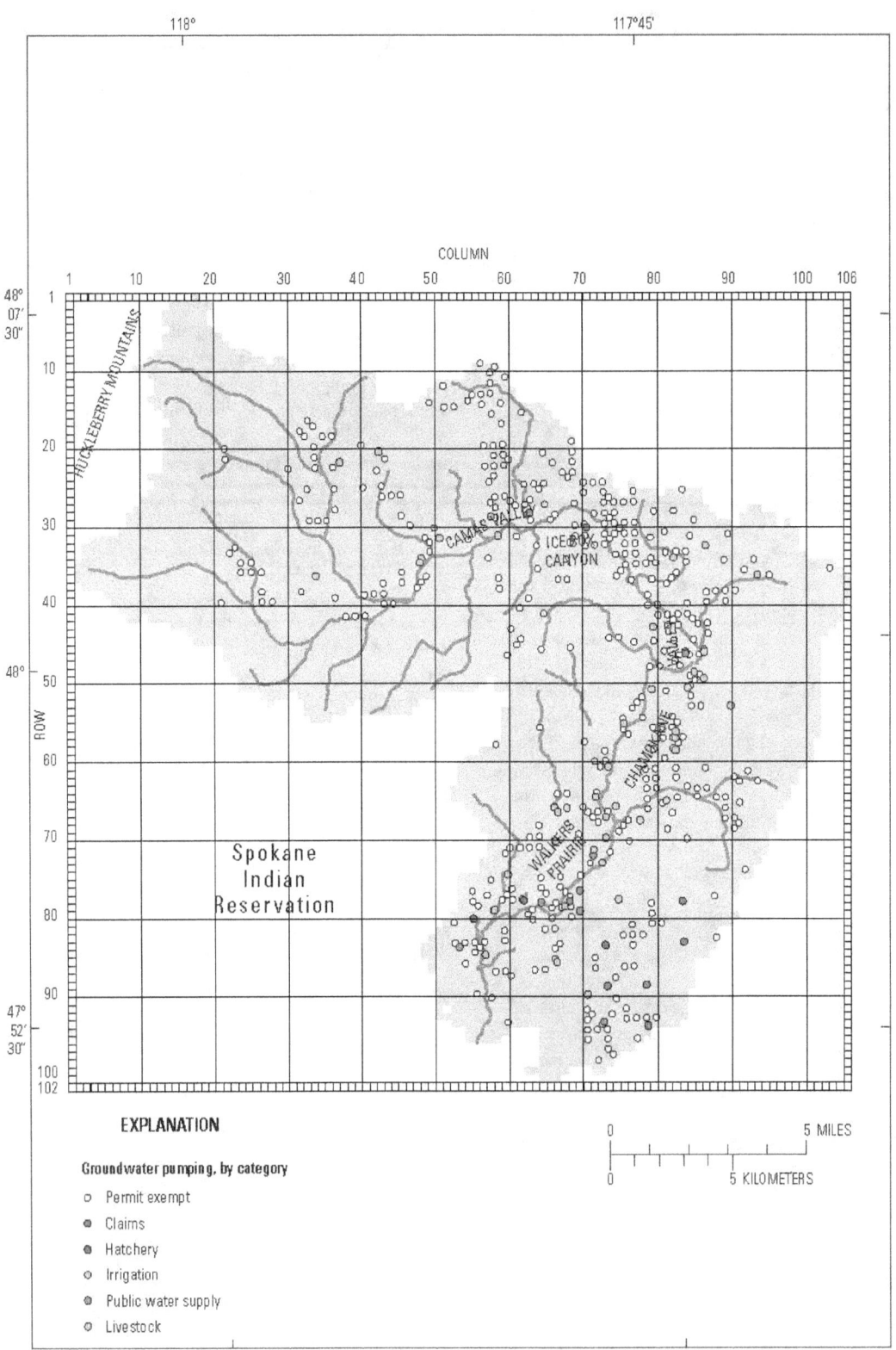

Figure 11.   Location and water-use categories of model wells, Chamokane Creek basin, Washington.

Table 2.   Simulated groundwater pumpage by stress period and category, Chamokane Creek basin, Washington.

[Values represent groundwater pumpage minus non-consumptive return flows]

Model stress period	Water year	Season	Groundwater pumpage, in acre-feet per year						
			Irrigation	Claims	Permit exempt	Public water supply	Hatchery	Livestock	Total
1	1980	October–March	0.0	−0.3	2.2	1.18	0.0	15.8	26.6
2	1980	April–September	1,250.0	1.7	22.7	1.18	0.0	13.2	1,296.4
3	1981	October–March	0.0	−0.3	2.2	1.18	0.0	15.8	26.6
4	1981	April–September	1,250.0	1.7	22.8	1.18	0.0	13.2	1,296.5
5	1982	October–March	0.0	−0.3	2.2	1.18	0.0	15.8	26.6
6	1982	April–September	1,250.0	1.7	23.0	1.18	0.0	13.2	1,296.7
7	1983	October–March	0.0	−0.3	2.3	1.18	0.0	15.8	26.6
8	1983	April–September	1,250.0	1.7	23.1	1.18	0.0	13.2	1,296.8
9	1984	October–March	0.0	−0.3	2.3	1.18	0.0	15.8	26.7
10	1984	April–September	1,250.0	1.7	23.3	1.18	0.0	13.2	1,297.0
11	1985	October–March	0.0	−0.3	2.3	1.18	0.0	15.8	26.6
12	1985	April–September	1,250.0	1.7	23.2	1.18	0.0	13.2	1,296.9
13	1986	October–March	0.0	−0.3	2.2	2.83	0.0	15.8	51.4
14	1986	April–September	1,250.0	1.7	22.8	4.49	0.0	13.2	1,321.2
15	1987	October–March	0.0	−0.3	2.2	4.56	0.0	15.8	52.5
16	1987	April–September	1,250.0	1.7	22.7	4.64	0.0	13.2	1,322.3
17	1988	October–March	0.0	−0.3	2.2	18.50	0.0	15.8	259.1
18	1988	April–September	1,250.0	1.7	22.2	32.35	0.0	13.2	1,528.5
19	1989	October–March	0.0	−0.3	2.1	33.04	0.0	15.8	270.0
20	1989	April–September	1,250.0	1.7	21.9	33.73	0.0	13.2	1,539.1
21	1990	October–March	0.0	−0.3	2.1	35.67	291.1	13.9	588.3
22	1990	April–September	1,250.0	1.7	21.7	37.62	588.7	13.8	2,157.4
23	1991	October–March	0.0	−0.3	2.2	38.70	588.7	13.4	901.6
24	1991	April–September	578.0	1.7	22.0	39.78	588.7	13.8	1,501.8
25	1992	October–March	0.0	−0.3	2.2	40.71	588.7	13.4	914.8
26	1992	April–September	578.0	1.7	23.0	41.63	588.7	13.8	1,515.9
27	1993	October–March	0.0	−0.3	2.3	42.96	588.7	13.4	942.6
28	1993	April–September	578.0	1.7	23.6	44.29	588.7	13.8	1,544.2
29	1994	October–March	0.0	−0.3	2.4	45.25	588.7	13.4	957.0
30	1994	April–September	578.0	1.7	24.8	46.20	588.7	13.8	1,559.7
31	1995	October–March	0.0	−0.3	2.5	47.67	588.7	13.6	979.2
32	1995	April–September	578.0	1.7	26.0	49.14	588.7	14.0	1,583.0
33	1996	October–March	0.0	−0.3	2.6	50.51	588.7	11.1	996.5
34	1996	April–September	578.0	1.7	27.1	51.89	588.7	11.4	1,601.2
35	1997	October–March	0.0	−0.3	2.8	53.27	676.5	11.7	1,106.5
36	1997	April–September	578.0	1.7	28.3	54.65	766.2	12.0	1,802.0
37	1998	October–March	0.0	−0.3	2.8	55.80	766.2	11.7	1,213.4
38	1998	April–September	578.0	1.7	28.6	56.94	766.2	12.0	1,819.5
39	1999	October–March	0.0	−0.3	2.8	58.14	766.2	11.7	1,231.4
40	1999	April–September	578.0	1.7	29.0	59.34	766.2	12.1	1,837.9
41	2000	October–March	0.0	−0.3	3.0	60.59	766.2	11.8	1,250.4
42	2000	April–September	578.0	1.7	30.5	61.84	766.2	12.2	1,858.2
43	2001	October–March	0.0	−0.3	3.0	63.16	770.4	12.0	1,274.4
44	2001	April–September	76.0	1.7	30.3	64.47	766.2	12.4	1,375.9
45	2002	October–March	0.0	−0.3	2.9	65.84	770.4	12.0	1,294.9
46	2002	April–September	76.0	1.7	30.0	67.22	766.2	12.4	1,396.1

Table 2.   Simulated groundwater pumpage by stress period and category, Chamokane Creek basin, Washington.—Continued

[Values represent groundwater pumpage minus non-consumptive return flows]

Model stress period	Water year	Season	Groundwater pumpage, in acre-feet per year						
			Irrigation	Claims	Permit exempt	Public water supply	Hatchery	Livestock	Total
47	2003	October–March	0.0	−0.3	2.9	68.66	1,665.1	12.4	2,211.5
48	2003	April–September	76.0	1.7	29.8	70.10	2,584.0	12.7	3,235.6
49	2004	October–March	0.0	−0.3	2.9	71.61	1,706.1	5.3	2,267.9
50	2004	April–September	76.0	1.7	29.4	73.11	3,430.9	5.6	4,097.6
51	2005	October–March	0.0	−0.3	2.9	74.69	1,971.3	5.5	2,557.0
52	2005	April–September	76.0	1.7	29.5	76.27	3,430.9	5.9	4,121.6
53	2006	October–March	0.0	−0.3	3.0	77.92	2,535.9	5.5	3,146.4
54	2006	April–September	76.0	1.7	30.3	79.58	3,995.4	5.9	4,711.6
55	2007	October–March	0.0	−0.3	3.0	81.31	2,535.9	5.5	3,172.3
56	2007	April–September	76.0	1.7	31.0	83.04	3,995.4	5.9	4,738.2
57	2008	October–March	0.0	−0.3	3.1	84.73	2,537.4	5.5	3,202.2
58	2008	April–September	456.0	1.7	31.8	86.43	3,997.0	5.9	5,148.8
59	2009	October–March	0.0	−0.3	3.2	88.43	2,537.4	5.5	3,231.8
60	2009	April–September	76.0	1.7	32.6	90.44	3,997.0	5.9	4,799.2
61	2010	October–March	0.0	−0.3	3.3	92.42	2,537.4	5.5	3,262.8
62	2010	April–September	76.0	1.7	33.4	94.40	3,997.0	5.9	4,830.8
		Average	333.4	0.7	14.5	44.8	1,040.5	11.7	1,745.5

Horizontal Hydraulic Conductivity

Horizontal isotropy was assumed for the basin-fill sediments, and each model layer was assigned one value for K_h, with the exception of model layers 1 and 2, which represent the Upper outwash aquifer. Simulated zones of hydraulic conductivity for the Upper outwash aquifer were based on mapped surficial hydrogeology (Kahle and others, 2010) and are shown in figure 12. The initial assigned values were estimated from K_h estimates for mapped units by Kahle and others (2010) and model-derived values from Ely and Kahle (2004). For the coarse-grained Upper outwash and Lower aquifers, initial K_h values ranged from 75 to 500 ft/d. The poorly sorted Landslide unit was assigned a K_h value of 6 ft/d and the fine-grained Valley confining unit was assigned a value of 10 ft/d. Initially, the Basalt and Bedrock units were assigned a K_h value of 0.3 and 0.05 ft/d, respectively, and those values were assigned to layers 7 and 8. During calibration, layer 8 was assigned an additional hydraulic property zone.

Vertical Hydraulic Conductivity

Vertical hydraulic conductivity (K_v) values were initially derived from ratios (vertical anisotropy) of the horizontal to the vertical values ($K_h{:}K_v$). Because of the unknown nature of, and local variations in, K_v, anisotropy ratios were regionalized using only two initial anisotropy ratios for the basin-fill units. The ratio for the Upper outwash aquifer, Landslide unit, and Lower aquifer was assumed to be 10:1. The Valley confining unit was assumed to be 100:1. The Basalt and Bedrock units were initially set to 1:1.

Storage Properties

Storage properties of the aquifers are known to be highly variable and there is a general lack of information for making reliable areal estimates. However, selected published values for similar aquifer and confining units were used as an initial estimate of storage properties. Both unconfined and confined conditions occur within the groundwater system. The specific storage value assigned to the sediments ranged from 0.05 to 0.0001 ft^{-1} and initial specific yield values ranged from 0.1 to 0.01. Model stability and convergence were highly sensitive to bedrock specific storage and specific yield values.

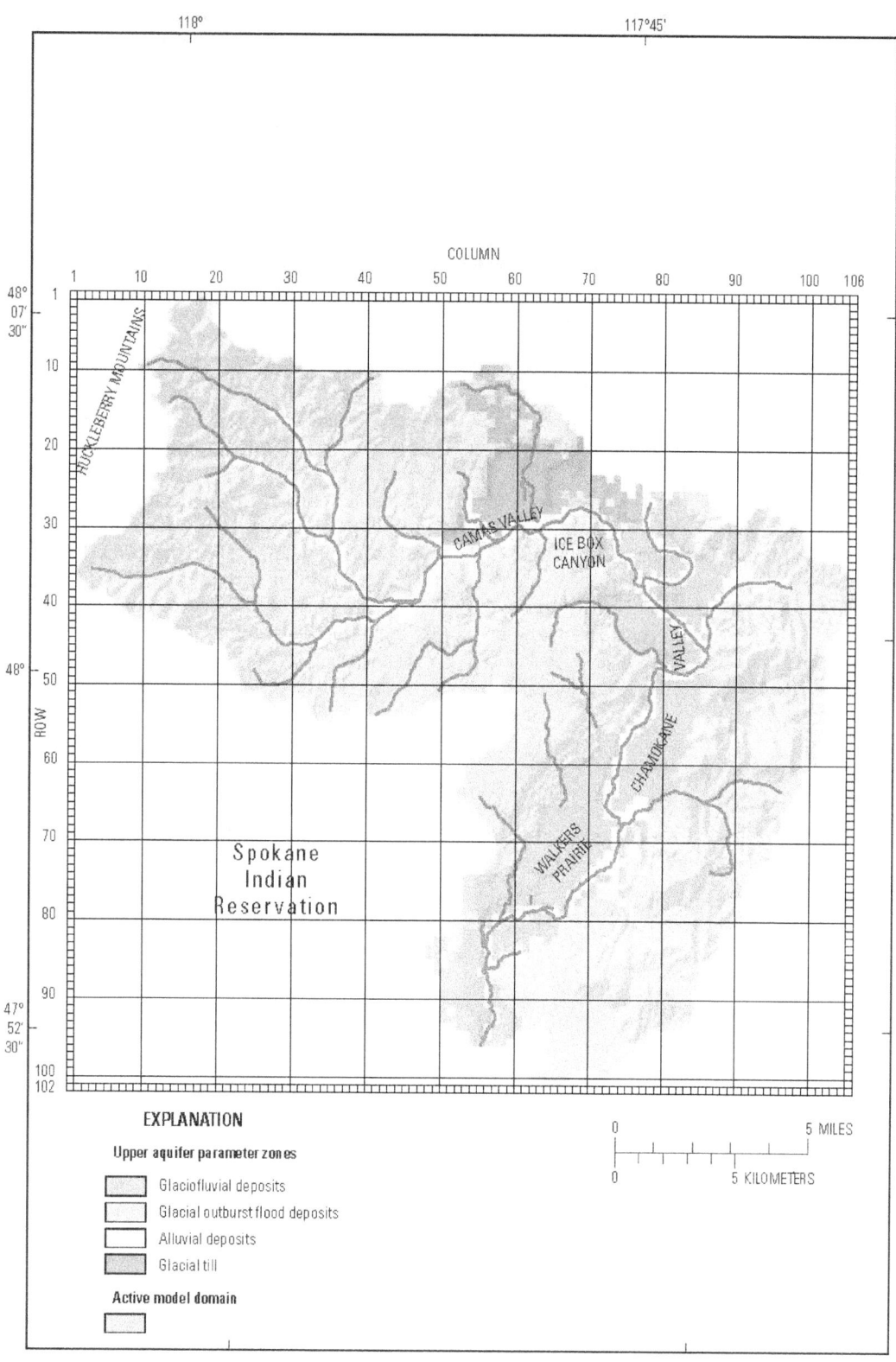

Figure 12. Simulated zones of horizontal hydraulic conductivity for model layers 1 and 2, Upper outwash aquifer, Chamokane Creek basin, Washington.

Description of Coupled Flow Model (GSFLOW)

GSFLOW simulates flow within and among three regions (fig. 13; Markstrom and others, 2008). The first region is bounded on top by the plant canopy and on the bottom by the lower limit of the soil zone; the second region consists of all streams and lakes; and the third region is the subsurface zone beneath the soil zone. PRMS is used to simulate hydrologic responses in the first region and MODFLOW is used to simulate hydrologic processes in the second and third regions.

Advantages of GSFLOW include the use of existing PRMS modules and MODFLOW packages, allowing for a flexible and adaptive design that incorporates both PRMS and MODFLOW frameworks. GSFLOW solves equations governing interdependent surface-water and groundwater flow using iterative solution techniques, creating the ability to simulate groundwater/surface-water flow in an integrated fashion.

The section 'Computation of Flow' in Markstrom and others (2008) provides a complete description of the computations made by GSFLOW, beginning with climate inputs of temperature, precipitation, and solar radiation, and ending with groundwater and its interactions with streams and lakes. Additional details on the computations of flow are described in the documentation for PRMS (Leavesley and others, 1983, 1996) and MODFLOW-2005 (Harbaugh, 2005).

Model Calibration

Model calibration is the process by which model parameters are adjusted to obtain a reasonable fit between simulated hydraulic heads and flows and measured data. Poorly quantified properties of the flow system can be constrained in the calibrated model on the basis of these measured water levels and streamflow. Throughout the calibration process, no adjustments were made that conflicted with the general understanding of the aquifer system and previously documented information.

Historical streamflow records are available for most of the simulation period (1988–2010) for one active USGS streamflow-gaging station, Chamokane Creek below Falls near Long Lake (12433200; fig. 1). Groundwater levels, some recorded by well drillers and others recorded by USGS personnel as part of the Phase 1 study, are available throughout the simulation period. Despite the existence of data for a longer time period, model calibration was limited to a 12-year period, water years 1999–2010. This period contained the most extensive dataset (a continuous streamflow record and the water levels measured as part of this study), represented a range of climatic conditions, and was sufficiently long to represent changes in groundwater conditions in response to climate variations.

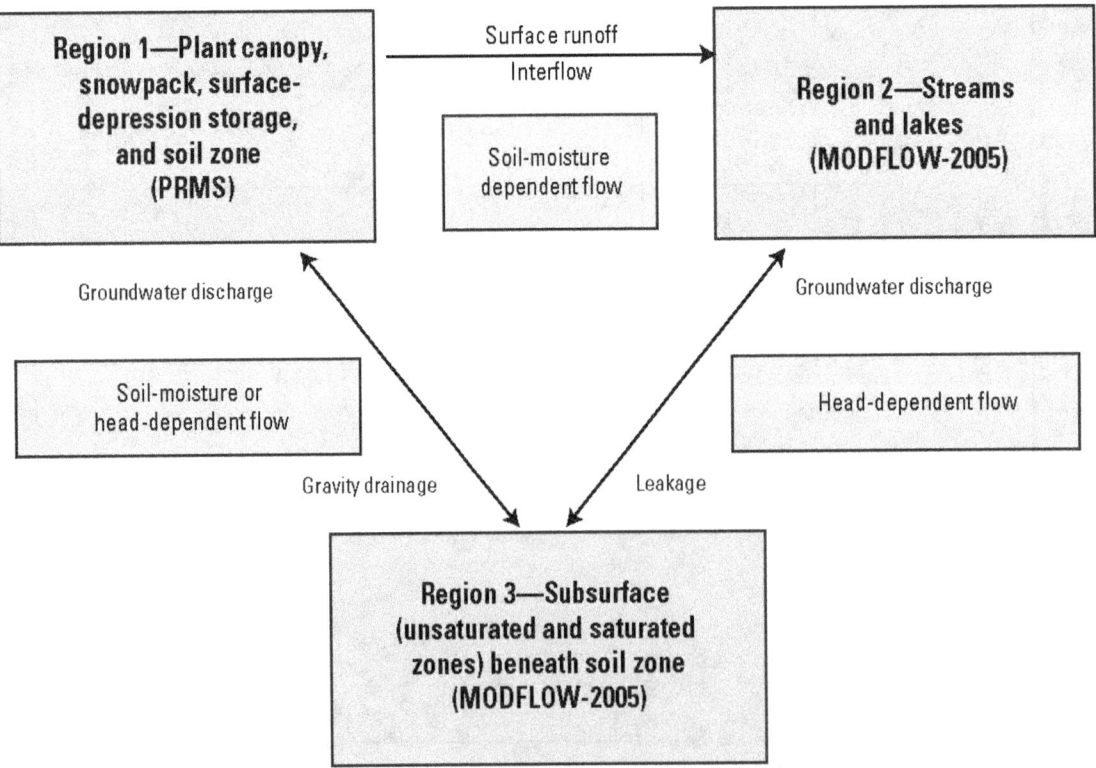

Figure 13.   The exchange of flow among the three regions in GSFLOW (from Markstrom and others, 2008).

Model calibration benefited from the subset of monthly water-level measurements collected as part of this study and streamflow measurements at the downstream outlet of the basin. The USGS National Water Information System database contained 108 wells with 739 water-level measurements (fig. 14) that were used in the calibration. Model observations are sparse in some areas and at some depths, and the paucity of data is most pronounced in the bedrock areas to the northwest and the eastern part of the model domain. Data for wells with water-level measurements over multiple months and years were available for analyzing temporal variations.

Calibration Approach

The model was calibrated using a combination of traditional trial-and-error modification of parameters and the automated parameter-estimation software package PEST (Doherty, 2010). PEST uses a nonlinear least-squares regression to find the set of parameter values that minimizes the weighted sum-of-squared-errors objective function.

MODFLOW parameters adjusted during calibration included K_h, K_h:K_v (vertical anisotropy), specific storage, specific yield, and streambed conductivity. PRMS and GSFLOW parameters adjusted during calibration were mainly those that distributed precipitation, air temperature, and evapotranspiration, and parameters used in the computation of flow into and out of the soil zone.

Model calibration was first conducted using a trial-and-error process to ensure that model predictions were in reasonable agreement with measured trends in groundwater levels and streamflow variations. Various attempts at automated calibration yielded valuable information, including identifying insensitive parameters. Some estimated parameter values were considered unreasonable or resulted in model instability. Each parameter optimization run was usually followed by another manual adjustment of parameter values to ensure good model fit with defensible parameterization schemes.

Automated calibrations of the estimated model parameters using PEST were conducted for the 12-year period (1999–2010) using 739 head observations and 4,383 daily streamflow observations.

Observations Weighting

Calibration using both measured groundwater levels and streamflows was done with observation weights adjusted to ensure equal contribution by the two groups to the model objective function, in accordance with the guidelines presented by Doherty and Hunt (2010). Observations and, therefore, residuals are weighted to allow a meaningful comparison of measurements with different units (weighted residuals are dimensionless) and to reduce the influence of measurements with large errors or uncertainty. The initial observation weight was defined using methods suggested by Hill (1998) and Hill and Tiedeman (2007). Errors in groundwater-level measurements were limited by the accuracy at wells whose locations were not measured using a GPS and by the accuracy of the DEM used to estimate the altitude. Errors in streamflow measurements generally are within 5 to 10 percent.

Model calibrations conducted using observations of different types require a weighting scheme that adequately represents the contribution to total model error of observations made in different measurement units. Modifications to the initial weighting approach were used to account for discrepancies in data density between the observation groups. A weighting scheme was designed to balance the contribution of the prevalence of daily mean streamflow measurements over groundwater levels. To redress this imbalance, weights for each class of observations were proportionally scaled such that water-level observations and streamflow observations each made roughly equivalent contributions to total model error. All water-level observations were assigned equal relative weights and streamflow observations were divided into low-flow (October 1–February 15 and June 1–September 30) and high-flow (February 16–May 31) periods and assigned an equal weight for each period. For the final parameter values, the contribution to the sum of squared weighted residuals of each of the three observation groups was approximately equal (low-flow observations, 36.4 percent; high-flow observations, 29.3 percent; and head observations, 34.3 percent).

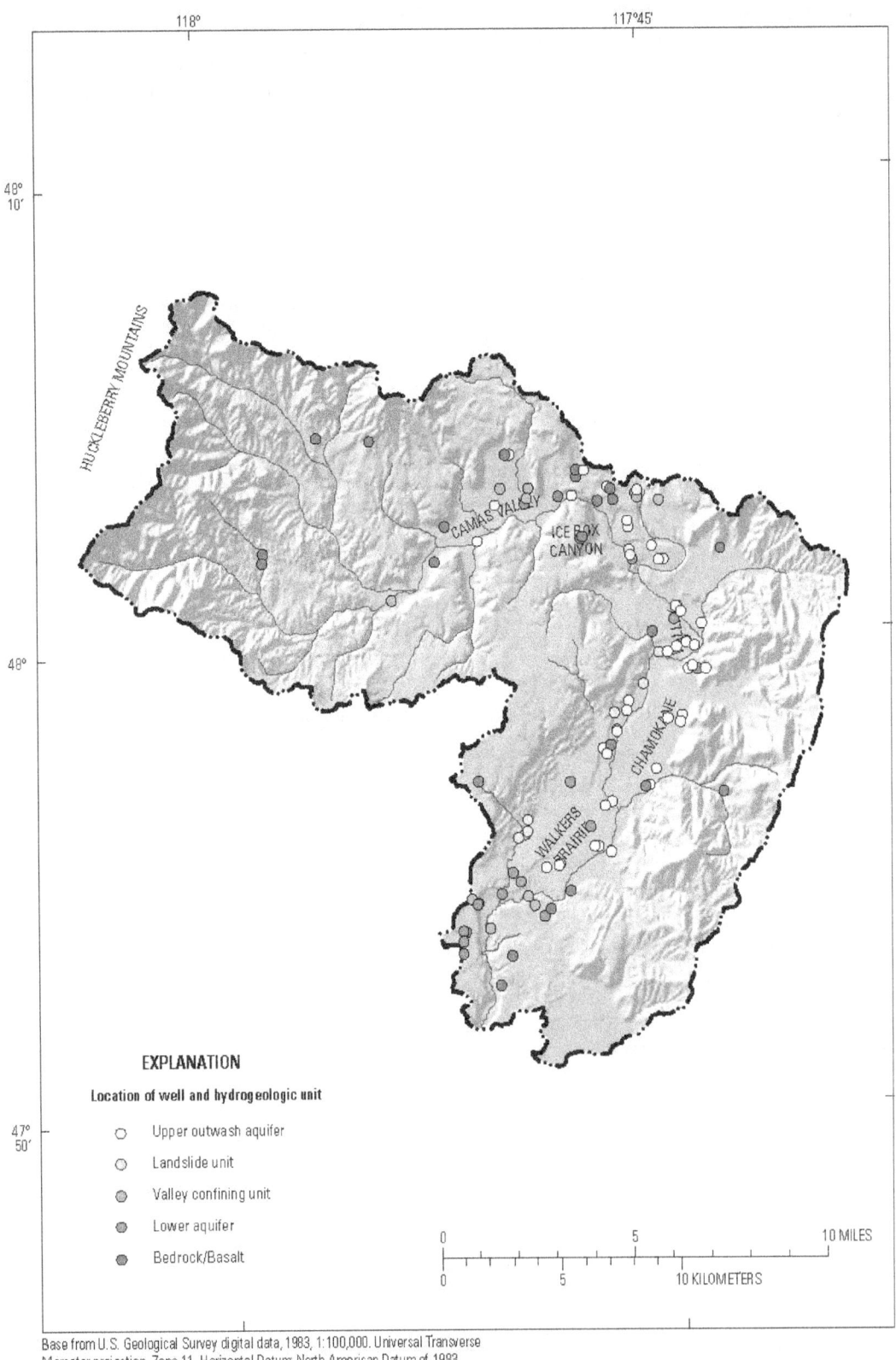

Base from U.S. Geological Survey digital data, 1983, 1:100,000. Universal Transverse
Mercator projection, Zone 11. Horizontal Datum: North American Datum of 1983

Figure 14.   Location of model hydraulic-head observations, Chamokane Creek basin, Washington.

Final Parameter Values and Sensitivities

PRMS Input

The PRMS parameter file was assembled using the GIS Weasel toolbox (Viger and Leavesley, 2007) and GSFLOW-specific parameters were added afterward. The Chamokane Creek model was delineated into 5,026 HRUs, corresponding directly to the active cells of the MODFLOW grid. Most parameters were not adjusted during calibration. Calibration focused on those parameters that controlled the distribution of precipitation and air temperature and flow in the soil zone. Measured data from the four climate stations (fig. 8; table 1) were used to develop the initial values of the temperature and precipitation lapse rate parameters specified in the PRMS parameter file. No changes were made to the x and y lapse rate parameters, but some adjustments were made to the z (altitude) lapse rate parameters. The available data from the relatively low-altitude climate stations did not represent the orographic effects on temperature and precipitation. Higher altitudes are cooler and wetter than the low-lying areas. The final calibrated values of selected parameters are shown in table 3.

MODFLOW Input

The final MODFLOW parameter values are shown in table 4. Hydraulic conductivity for the Upper outwash aquifer, which represents mostly glacial outwash sand and gravel, ranged from 100 to 345 ft/d. The lower conductivity was for the low-permeability glacial till to the north (UA_ti; fig. 12). Simulated hydraulic conductivity for the Landslide unit, which represents poorly sorted broken basalts and sedimentary interbeds, was estimated to be 22 ft/d. Estimated hydraulic conductivity for the Valley confining unit, which represents a thick, low-permeability unit consisting mostly of extensive glaciolacustrine silt and clay, was estimated to be 4 ft/d. Estimated hydraulic conductivity for the Lower aquifer, which represents mostly sand and some gravel, was estimated to be 135 ft/d. This final parameter value for the Lower aquifer is higher than the median value reported in Kahle and others (2010) but falls within the reported range. During model calibration, the Bedrock unit, and to a lesser degree, the Basalt unit was very important to the simulation of streamflow. Spatially, the Basalt and Bedrock units exist at depth throughout the entire model domain and most of the precipitation and snowmelt occurs at high altitudes where these units are at the surface. For these reasons, a third hydraulic parameter zone was added for all of layer 8. This approach allowed the upper Bedrock and Basalt units (layer 7) and lower Bedrock unit to be estimated separately, but final parameter values for all three are similar. Hydraulic conductivity for the upper Bedrock, Basalt, and lower Bedrock units was estimated to be 0.5, 0.5, and 0.32 ft/d, respectively.

Vertical anisotropy, the ratio of horizontal to vertical hydraulic conductivity, was assigned a value of 10:1 for the Upper outwash aquifer and Lower aquifer. Simulation results were insensitive to these parameters, so no estimation was done. The vertical anisotropy for the Landslide unit and Valley confining unit were estimated to be 10:1 and 100:1, respectively. The regression was very sensitive to the vertical anisotropy of the Bedrock and Basalt units for the reasons mentioned above and some model instability was introduced by the adjustment of these parameters. Both the PEST and trial-and-error calibration methods were used and the values were finally set to 10:1 for the upper Bedrock and Basalt units and 100:1 for the lower Bedrock unit.

Simulation results were sensitive to storage properties, and small adjustments to the bedrock specific storage and specific yield caused model instability. All storage properties were included in the parameter estimation, but after some initial adjustments, were set to their final values. Specific storage values fall within acceptable ranges, but specific yields for the Bedrock and Basalt units, and to a lesser degree Landslide unit and Valley confining unit, are lower than expected. The final storage values provided model stability and produced acceptable simulated water levels.

Streambed conductance was based on computed stream depth and width, an assigned streambed thickness of 1 ft, and an estimated streambed hydraulic conductivity. Streambed hydraulic conductivity was initially assigned a value of 1 ft/d. During the calibration phase, however, the streams were grouped into three broad categories and hydraulic conductivities were adjusted to achieve a good fit between simulated and measured gains and losses. Streambed conductivity values of tributary reaches that flowed over the Bedrock and Basalt units (SFR_TRIB) was estimated to be 0.5 ft/d. Streambed conductivity values of upper Chamokane Creek (North, South, and Middle Forks; SFR_FORK) was estimated to be 0.7 ft/d. Mainstem streambed conductivity values of Chamokane Creek (SFR_CHAM), including all of Chamokane Creek downstream of Ice Box Canyon, was estimated to be 0.9 ft/d.

Sensitivities

Sensitivity analysis is used to assess the effects of different conceptual models (different model designs and parameter values) on the simulated heads and flows, and to develop useful nonlinear regressions (Hill, 1998; Hill and Tiedeman, 2003). The ability to estimate a parameter value using nonlinear regression is a function of the sensitivity of simulated values such as groundwater levels and streamflow to changes in the parameter value. Parameter sensitivity reflects the amount of information about a parameter that is provided by the observation data. Generally speaking, if a parameter has a high sensitivity, observation data exist to effectively estimate the parameter value. If the parameter has low sensitivity, observation data are not sufficient to estimate the parameter value and changing the parameter value will have little effect on the sum of squared errors.

Table 3.   Values and source of non-default PRMS parameters used for the model of Chamokane Creek basin, Washington.

[**Source:** C, parameters that cannot be estimated from available data and are adjusted during calibration; CG, parameters that are initially computed in GIS and are adjusted, preserving relative spatial variation during calibration; L, parameters obtained from the literature as estimated or empirical estimates]

Parameter	Description	Minimum value	Maximum value	Source
adjmix_rain	Monthly adjustment factor for a mixed precipitation event as a decimal fraction	0.00016	0.68	L
covden_sum	Summer plant canopy density as a decimal fraction of the HRU area	0.19	1.00	CG
covden_win	Winter plant canopy density as a decimal fraction of the HRU area	0.00	1.00	CG
dday_intcp	Intercept of monthly degree-day to temperature relation	−60.95	9.93	L
dday_slope	Slope of monthly degree-day to temperature relation	0.44	0.81	L
fastcoef_lin	Linear flow-routing coefficient for fast interflow	0.048	0.048	C
fastcoef_sq	Non-linear flow-routing coefficient for fast interflow	0.80	0.80	C
gwflow_coef	Linear coefficient to route water in groundwater reservoir to streams	0.010	0.010	C
jh_coef	Monthly air temperature coefficient used in Jensen-Haise potential evapotranspiration equation	0.011	0.046	C
jh_coef_hru	Air temperature coefficient used in Jensen-Haise potential evapotranspiration equation for each HRU	14.14	16.47	CG
potet_sublim	Fraction of potential evapotranspiration sublimated from snow surface as a decimal fraction	0.41	0.41	C
pref_flow_den	Decimal fraction of the soil zone available for preferential flow	0.01	0.01	C
rad_trncf	Transmission coefficient for short-wave radiation through winter plant canopy on an HRU as a decimal fraction	0.063	0.99	CG
sat_threshold	Maximum volume of water per unit area in the soil zone	10.00	10.00	C
slowcoef_lin	Linear flow-routing coefficient for slow interflow	0.080	0.080	C
slowcoef_sq	Non-linear flow-routing coefficient for slow interflow	0.043	0.043	C
smidx_coef	Coefficient in non-linear contributing area algorithm	0.0010	0.0010	C
snarea_curve	Snow area-depletion curve values, for each curve as a decimal fraction	0.050	1.00	L
snarea_thresh	Maximum water equivalent threshold, water equivalent in an HRU less than threshold results in use of snow-covered-area curv	0.0040	14.60	CG
snowinfil_max	Daily maximum snowmelt infiltration for the HRU	2.50	2.50	C
soil_moist_max	Maximum volume of water per unit area in the capillary reservoir	1.81	12.89	CG
soil_rechr_max	Maximum value in capillary reservoir where evaporation and transpiration can occur simultaneously	0.60	1.61	CG
srain_intcp	Maximum summer rain storage in the plant canopy for plant type on HRU	0.00	0.050	L
ssr2gw_exp	Exponent in the equation used to compute gravity drainage to PRMS groundwater reservoir or MODFLOW finite-difference cell	0.16	0.16	C
ssr2gw_rate	Linear coefficient in the equation used to compute gravity drainage to PRMS groundwater reservoir or MODFLOW finite-difference cell	0.26	0.26	C
tmax_allrain	Monthly minimum air temperature at an HRU that results in all precipitation during a day being rain	60.00	60.00	L
tmax_allsnow	Monthly maximum air temperature at which precipitation is all snow for the HRU	38.00	38.00	L
transp_beg	Begin month for transpiration computations at HRU	4.00	4.00	C
transp_end	Last month for transpiration computations at HRU	10.00	10.00	C
wrain_intcp	Maximum winter rain storage in the plant canopy for plant type on HRU	0.00	0.050	L

Table 4. Final calibrated MODFLOW parameters used for the model of Chamokane Creek basin, Washington.

[ft/d, feet per day; K_h, horizontal hydraulic conductivity; K_v, vertical hydraulic conductivity]

Parameter description	Parameter abbreviation	Horizontal hydraulic conductivity (ft/d)	Vertical anisotropy $(K_h{:}K_v)$	Specific storage (ft^{-1})	Specific yield
Glacial till	UA_ti	100	10	5.0×10^{-4}	1.0×10^{-1}
Alluvial deposits	UA_al	345	10	5.0×10^{-4}	1.0×10^{-1}
Glacial outburst flood deposits	UA_f	309	10	5.0×10^{-4}	1.0×10^{-1}
Glaciofluvial deposits	UA_gf	309	10	5.0×10^{-4}	1.0×10^{-1}
Landslide unit	LU	22	10	5.0×10^{-4}	1.0×10^{-2}
Valley confining unit	VC	4	100	5.0×10^{-5}	1.0×10^{-2}
Lower aquifer	LA	135	10	5.0×10^{-4}	1.0×10^{-1}
Basalt unit	BT	0.5	10	1.0×10^{-6}	2.5×10^{-4}
Upper bedrock unit	BK	0.5	10	1.0×10^{-6}	2.5×10^{-4}
Lower bedrock unit	LBK	0.32	100	1.0×10^{-6}	

Parameter description	Parameter abbreviation	Streambed conductivity (ft/d)
Streambed conductivity values of tributary reaches that flowed over the Bedrock and Basalt units	SFR_TRIB	0.5
Streambed conductivity values for upper Chamokane Creek (North, South, and Middle Forks)	SFR_FORK	0.7
Streambed conductivity values of the mainstem of Chamokane Creek	SFR_CHAM	0.9

Normalized composite scaled sensitivities (CSS) for 2 "average" water years (2007–08) are shown in figure 15. CSS reflect the total amount of information provided by the observations for the estimation of one parameter. Presenting CSS for 2007–08 show the relative importance of groundwater-level and streamflow observations to model parameters under average climatic conditions. Groundwater-level and streamflow observations are most sensitive to parameters used for the computation of fast interflow from preferential-flow reservoirs (FASTCOEF_LIN), hydraulic conductivity of the lower bedrock (KX_LBK; model layer 8), and streambed conductivity of upper Chamokane Creek (SFR_FORK). Assessing parameter sensitivities with respect to the separate observation groups is more instructive (table 5). Low-flow observations (October 1–February 15; June 1–September 30) were most affected by the hydraulic conductivity of the glaciofluvial deposits (KX_UAgf) that composes much of the upper outwash aquifer along the Chamokane Valley.

Those sediments largely control groundwater flow in the Upper outwash aquifer and toward Chamokane Creek. High-flow observations (February 16–May 31) were most affected by fast interflow from preferential-flow reservoirs (FASTCOEF_LIN) and the hydraulic conductivity of bedrock (KX_LBK). Streamflow during this time period is mostly snowmelt and precipitation events on snow, so parameters that control the infiltration of precipitation, snowmelt, and Hortonian runoff (Horton, 1933) in the upper drainage basin should be of greater importance. Hydraulic-head observation sensitivities were greatest for horizontal and vertical hydraulic conductivity of the lower bedrock (KX_LBK and KV_LBK, respectively), and streambed conductivity of the tributary streams to Chamokane Creek (SFR_TRIB). The normalized composite scaled sensitivity for the specific yield of the Lower aquifer (SY_LA) is 0, implying the unit remains confined throughout the simulation.

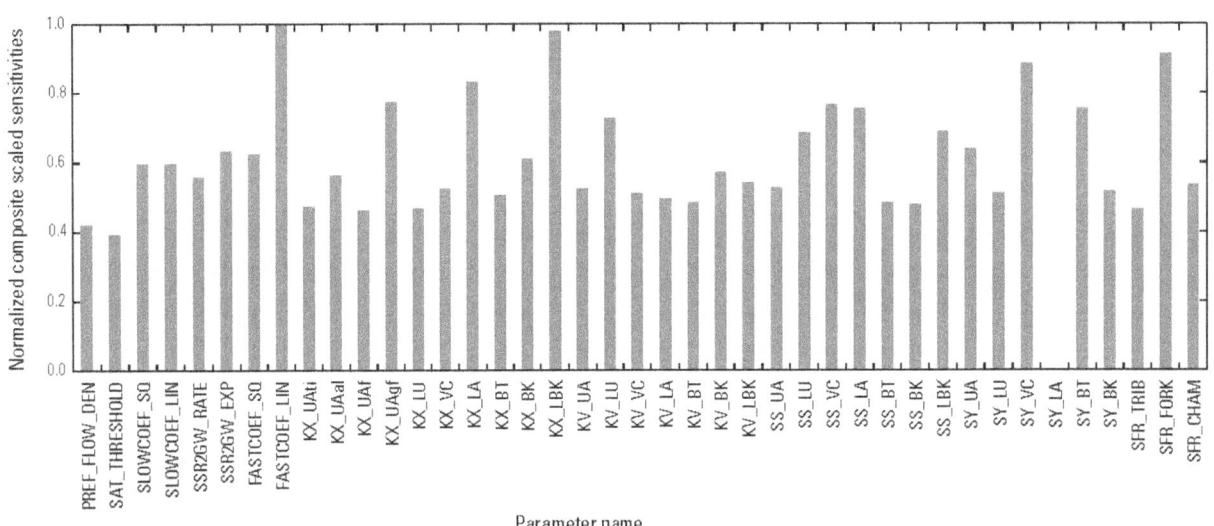

Figure 15. Normalized composite scaled sensitivities of final calibrated model parameters to hydraulic-head and streamflow observations, Chamokane Creek basin, Washington. Descriptions of parameter names are shown in tables 3 and 4.

Table 5. Normalized composite scaled sensitivities of final calibrated model parameters to hydraulic-head and streamflow observations, Chamokane Creek basin, Washington.

[**Parameter group:** kx, hydraulic conductivity; kv, vertical anisotropy; ss, specific storage; sy, specific yield; sfr, streamflow-routing. **Parameter name:** Descriptions are shown in tables 3 and 4]

Parameter group	Parameter name	Model observations			
		Low flow	High flow	Head	All
soil	PREF_FLOW_DEN	0.594	0.228	0.304	0.419
	SAT_THRESHOLD	0.528	0.224	0.482	0.392
	SLOWCOEF_SQ	0.585	0.525	0.471	0.598
	SLOWCOEF_LIN	0.548	0.534	0.588	0.597
	SSR2GW_RATE	0.696	0.379	0.622	0.558
	SSR2GW_EXP	0.702	0.495	0.711	0.634
	FASTCOEF_SQ	0.618	0.519	0.836	0.625
	FASTCOEF_LIN	0.710	1.000	0.443	1.000
kx	KX_UAti	0.623	0.279	0.642	0.474
	KX_UAal	0.819	0.272	0.531	0.564
	KX_UAf	0.600	0.253	0.777	0.461
	KX_UAgf	1.000	0.504	0.778	0.774
	KX_LU	0.535	0.354	0.513	0.467
	KX_VC	0.705	0.315	0.570	0.526
	KX_LA	0.805	0.736	0.674	0.834
	KX_BT	0.708	0.243	0.727	0.506
	KX_BK	0.490	0.582	0.532	0.612
	KX_LBK	0.677	0.967	0.894	0.980
kv	KV_UA	0.711	0.314	0.496	0.525
	KV_LU	0.623	0.679	0.570	0.727
	KV_VC	0.614	0.362	0.615	0.512
	KV_LA	0.583	0.352	0.670	0.496
	KV_BT	0.637	0.277	0.697	0.484
	KV_BK	0.652	0.441	0.537	0.571
	KV_LBK	0.731	0.252	0.984	0.542
ss	SS_UA	0.657	0.360	0.594	0.528
	SS_LU	0.509	0.677	0.375	0.686
	SS_VC	0.648	0.707	0.783	0.766
	SS_LA	0.616	0.717	0.597	0.756
	SS_BT	0.661	0.270	0.547	0.483
	SS_BK	0.642	0.263	0.700	0.480
	SS_LBK	0.516	0.660	0.746	0.690
sy	SY_UA	0.847	0.393	0.695	0.640
	SY_LU	0.668	0.323	0.560	0.513
	SY_VC	0.577	0.893	0.662	0.886
	SY_LA	0.000	0.000	0.000	0.000
	SY_BT	0.570	0.730	0.694	0.756
	SY_BK	0.574	0.409	0.500	0.517
sfr	SFR_TRIB	0.543	0.269	1.000	0.463
	SFR_FORK	0.650	0.904	0.630	0.913
	SFR_CHAM	0.749	0.272	0.666	0.535

Assessment of Model Fit

A graphical and descriptive comparison of simulated and measured groundwater levels and streamflow values provides a clear insight to the model fit and complements the statistical measures of model fit. Such a comparison indicates how well the model replicates the flow system. Although the error variance for the model is well within an acceptable limit, it is important to determine that the model accurately simulates the regional direction and amounts of flow in the groundwater-flow system (directions and amounts of flow).

Comparison of Measured and Simulated Hydraulic Heads

A traditional and intuitive assessment of model calibration is a simple plot of measured hydraulic heads as a function of simulated hydraulic heads (fig. 16). At 108 well measurement points, the mean and median difference between 739 simulated and measured hydraulic heads are 7 and 11 ft, respectively. The residuals for the 12-year simulation period show that 69 percent of the simulated heads exceeded measured heads with a median residual value of 19 ft, and 31 percent were less than measured heads with a median residual value of -16 ft. These results indicate that there was some bias in the calibrated model toward overpredicting groundwater levels at the measurement points. The root-mean-square (RMS) error of the difference between simulated and

measured hydraulic heads in the observation wells, divided by the total difference in water levels in the groundwater system (Anderson and Woessner, 1992, p. 241), also should be less than 10 percent to be acceptable (Drost and others, 1999). The calibrated model produces an RMS error divided by the total difference in water levels of 4.7 percent.

The spatial distribution of the residuals (that is, differences between the simulated and measured groundwater levels) reflects the general bias in the model results (fig. 17). For example, high heads were simulated upstream (west) of the inlet of Ice Box Canyon and low heads were simulated downstream (east) of the outlet of Ice Box Canyon. The largest negative residuals were simulated in the bedrock and basalt units (model layers 7 and 8) suggesting that the relatively simple representation of these hydrogeologic units does not adequately represent their heterogeneity. The model tends to overpredict hydraulic heads (simulated heads are greater than measured heads) and underpredict heads (simulated heads are less than measured heads) along the upper parts of the Chamokane Valley toward the basin outlet.

For simulated hydraulic heads to be acceptable, the distribution of heads and the patterns of flow also should approximate the generalized water-level distributions and flow patterns mapped as part of this study (figs. 5 and 6). Simulated groundwater-flow patterns in the Upper outwash aquifer and Lower aquifer generally match the mapped patterns; the simulated water-level contours have similar shape and thus, flow directions are similar (fig. 18).

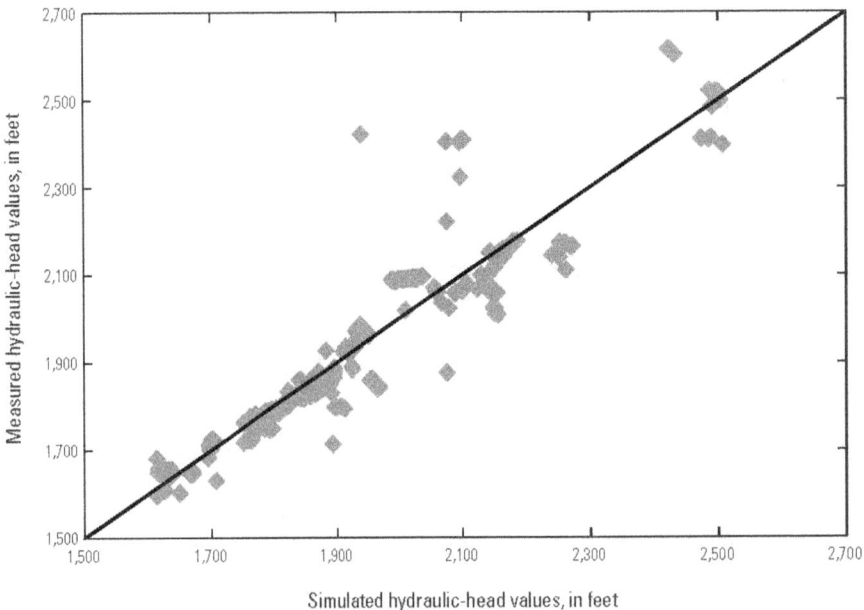

Figure 16. Measured hydraulic heads as a function of simulated hydraulic heads, Chamokane Creek basin, Washington.

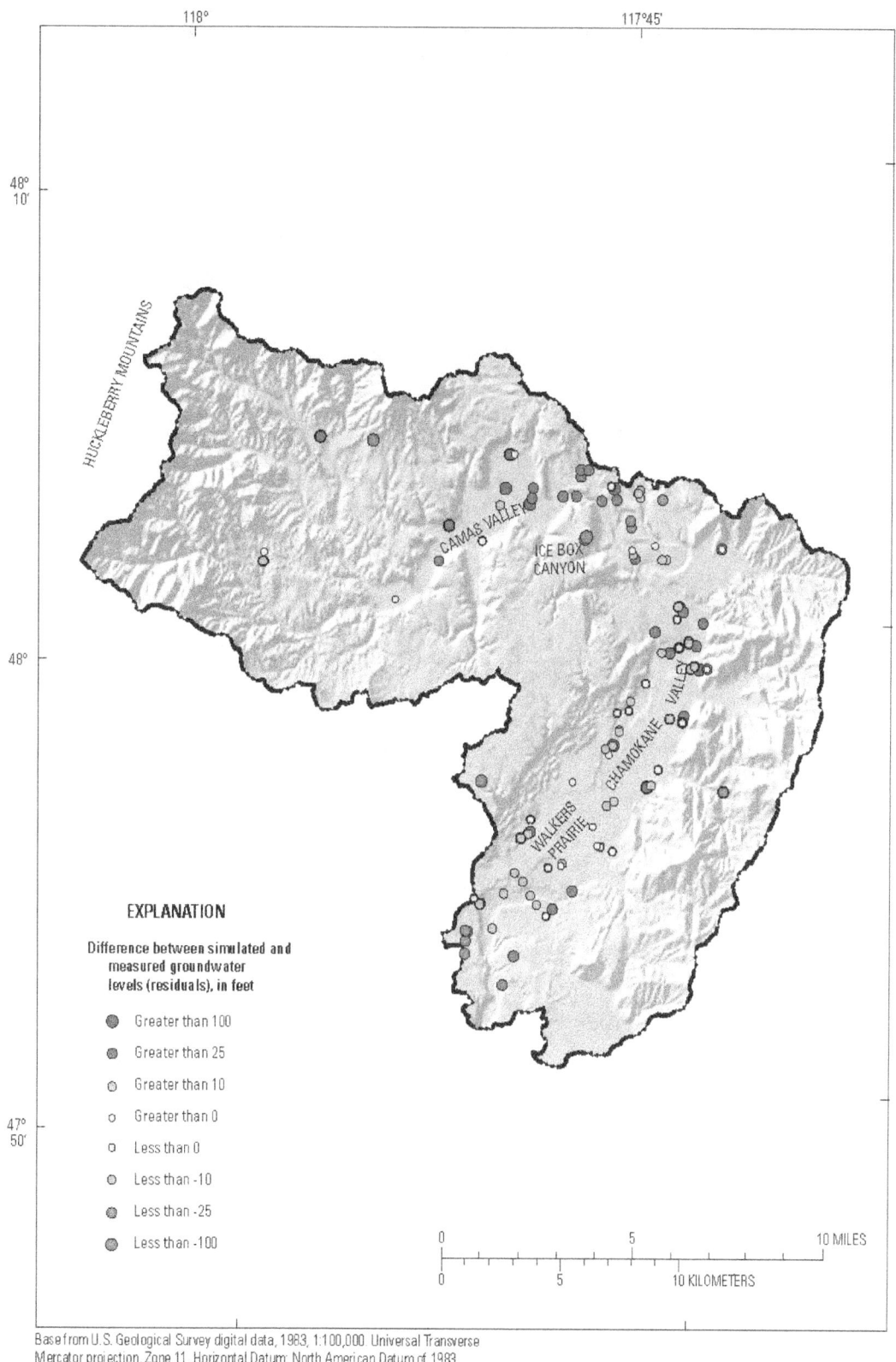

EXPLANATION

Difference between simulated and
measured groundwater
levels (residuals), in feet

- Greater than 100
- Greater than 25
- Greater than 10
- Greater than 0
- Less than 0
- Less than -10
- Less than -25
- Less than -100

Base from U.S. Geological Survey digital data, 1983, 1:100,000. Universal Transverse
Mercator projection, Zone 11. Horizontal Datum: North American Datum of 1983

Figure 17. Differences between simulated and measured groundwater levels (residuals), Chamokane
Creek basin, Washington.

A. **Upper outwash aquifer**

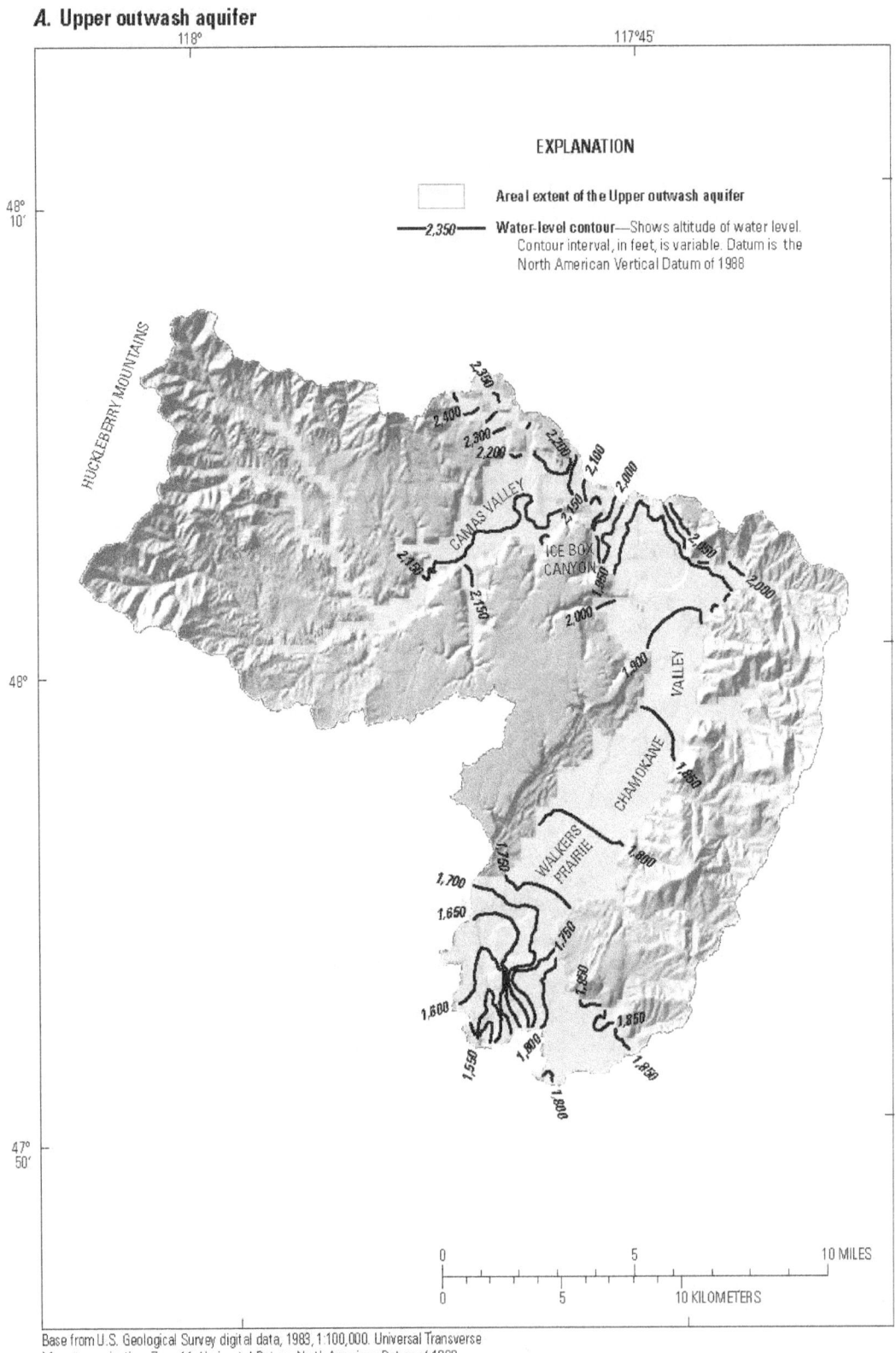

Base from U.S. Geological Survey digital data, 1983, 1:100,000. Universal Transverse
Mercator projection, Zone 11. Horizontal Datum: North American Datum of 1983

Figure 18.   Areal extent and simulated water-level altitudes in the (*A*) Upper outwash aquifer and (*B*) Lower aquifer, Chamokane Creek basin, Stevens County, Washington.

B. Lower aquifer

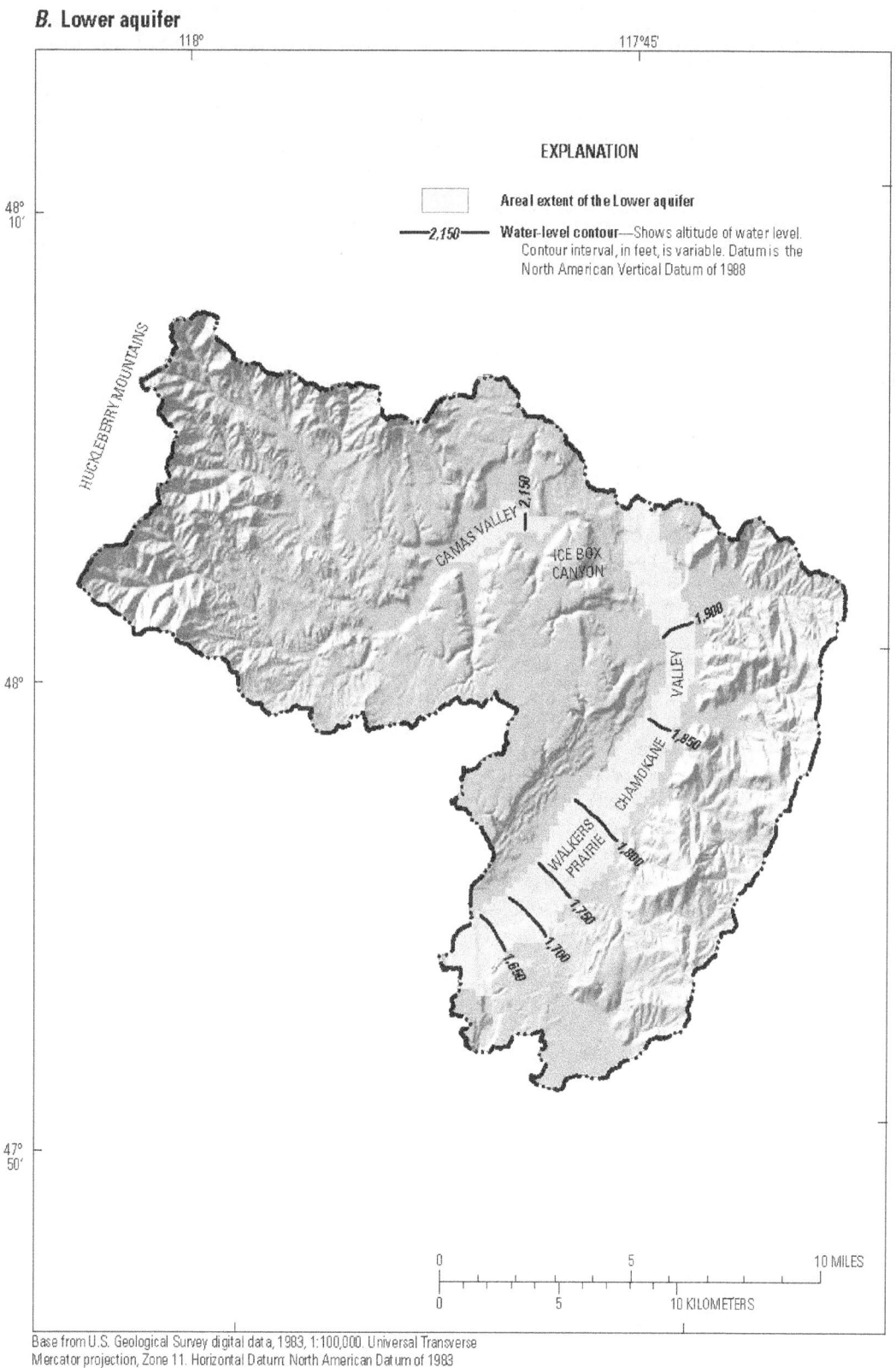

Base from U.S. Geological Survey digital data, 1983, 1:100,000. Universal Transverse
Mercator projection, Zone 11. Horizontal Datum: North American Datum of 1983

Figure 18.—Continued

A subset of simulated and measured monthly water-level measurements collected as part of this study is presented to assess the model's ability to simulate temporal variations (figs. 19 and 20A–20G). In general, the model is able to reproduce the temporal patterns and vertical gradients of the measured groundwater hydrographs. In almost all instances presented here, the simulated water levels are greater than the measured water levels. The one exception is well 28N/39E-26E01 (fig. 20G), which is completed in the Lower aquifer and located near the basin outlet. This follows the general model pattern of water-level overprediction in the upper basin and water-level underprediction lower in the basin. Simulated and measured hydrographs for wells 30N/39E-25Q02 (fig. 20A) and 29N/40E-23M06 (fig. 20C), both completed in the Upper outwash aquifer, show the model's ability to reproduce the annual pattern of water-level rises in the late spring to early summer followed by a decline from late summer through early spring. The model did a poor job reproducing the measured water levels at well 29N/40E-22P01 (fig. 20D), completed in the Landslide unit.

Three sets of paired wells are presented here to examine the model's ability to simulate the measured vertical gradients. Water levels at well pairs 29N/40E-15R02 and 29N/40E-15Q02 (fig. 20B) and 28N/40E-05A01 and 28N/40E-05A02 (fig. 20E) have a measured and simulated upward gradient. Water levels at wells 28N/40E-17J01 and 28N/40E-17C01 (fig. 20F) have a measured and simulated downward gradient. Although simulated water levels at wells 29N/40E-15R02 and 29N/40E-15Q02 (fig. 20B) do reproduce the observed upward gradient, the model does not capture the seasonal trend of rising water levels during the late spring and early summer. Simulated water levels at these locations remain relatively flat. The non-response could be due in part to their proximity to a streamflow boundary condition or an error in the estimated storage property.

Comparison of Measured and Simulated Streamflow

Simulated and measured streamflow have many similarities and generally show close agreement, especially during the late summer and early autumn baseflow period, as demonstrated by hydrographs of simulated and measured daily mean streamflows for the streamflow-gaging station for water years 2000–2009 (fig. 21). Simulated and measured streamflow show poor agreement for 2010. Water year 2010 was an unusual year, not in total amount of streamflow, but in the pattern of runoff. Unlike most years with a pronounced period of snowmelt followed by a recessionary limb of the hydrograph, 2010 had multiple smaller peaks followed by periods of low flow. Precipitation events likely followed different patterns and lapse rates that were not captured by the model.

Simulated and measured mean monthly streamflow for the streamflow-gaging station, which includes spring runoff of snowmelt and autumn/winter baseflows, demonstrate good agreement between the two (fig. 22). In general, the model tends to underpredict streamflow (simulated value less than measured value) from January to June and overpredict streamflow (simulated value greater than measured value) from July to December. The largest errors occur because the model failed to simulate large, but short term (1–2 day) January snowmelt events. Simulated and measured annual mean streamflow for the streamflow-gaging station is shown in figure 23. Annual differences between simulated and measured streamflow for the site ranged from -63 to 22 percent.

Another useful way to examine "goodness of fit" between simulated and measured streamflow is by calculating the volumetric efficiency (VE) proposed by Criss and Winston (2008).

$$VE = 1 - \frac{\sum |Qcalc - Qobs|}{\sum Qobs} \qquad (1)$$

where

    $Qobs$ is the observed streamflow and
    $Qcalc$ is the simulated or prediced streamflow.

The VE thus ranges from 0 to 1 and represents the fractional volumetric mismatch between the measured and simulated values (Criss and Winston, 2008). For an unbiased model, VE = 1 and the predicted total volume of water delivered over a given time interval matches the actual volume delivered. For the 12-year calibration period, the VE = 0.73. More instructive are the VEs for each water year and the average VE for each month (table 6). Four of the years had a VE greater than 0.8 and 7 years had a VE greater than 0.7. As mentioned earlier, simulated and measured streamflow for 2010 did not show close agreement. Average monthly VEs were highest for July–December, and all but March had a VE greater than 0.7. Water-resource management issues in the Chamokane Creek basin tend to occur in the late summer to early autumn, so the model's ability to reproduce monthly streamflows during this time was important.

Although total simulated streamflow at the streamflow-gaging station matched measured streamflow reasonably well, the model did not reproduce the pattern of streamflow gains. In the two low-flow seepage investigations completed during Phase 1 of the study, measured streamflows at site 12433175 (Chamokane Creek at Ford-Wellpinit Road, near Ford) were 1.26 and 1.65 ft³/s (Kahle and others, 2010). On the same or next day, measured streamflows at the USGS streamflow-gaging station were 28 and 29.6 ft³/s. These measurements indicate about 95 percent of total streamflow at the gage was gained downstream of Ford to near the gaging station. In contrast, the model simulated only about 65 percent of total streamflow was gained downstream of Ford to the gaging station.

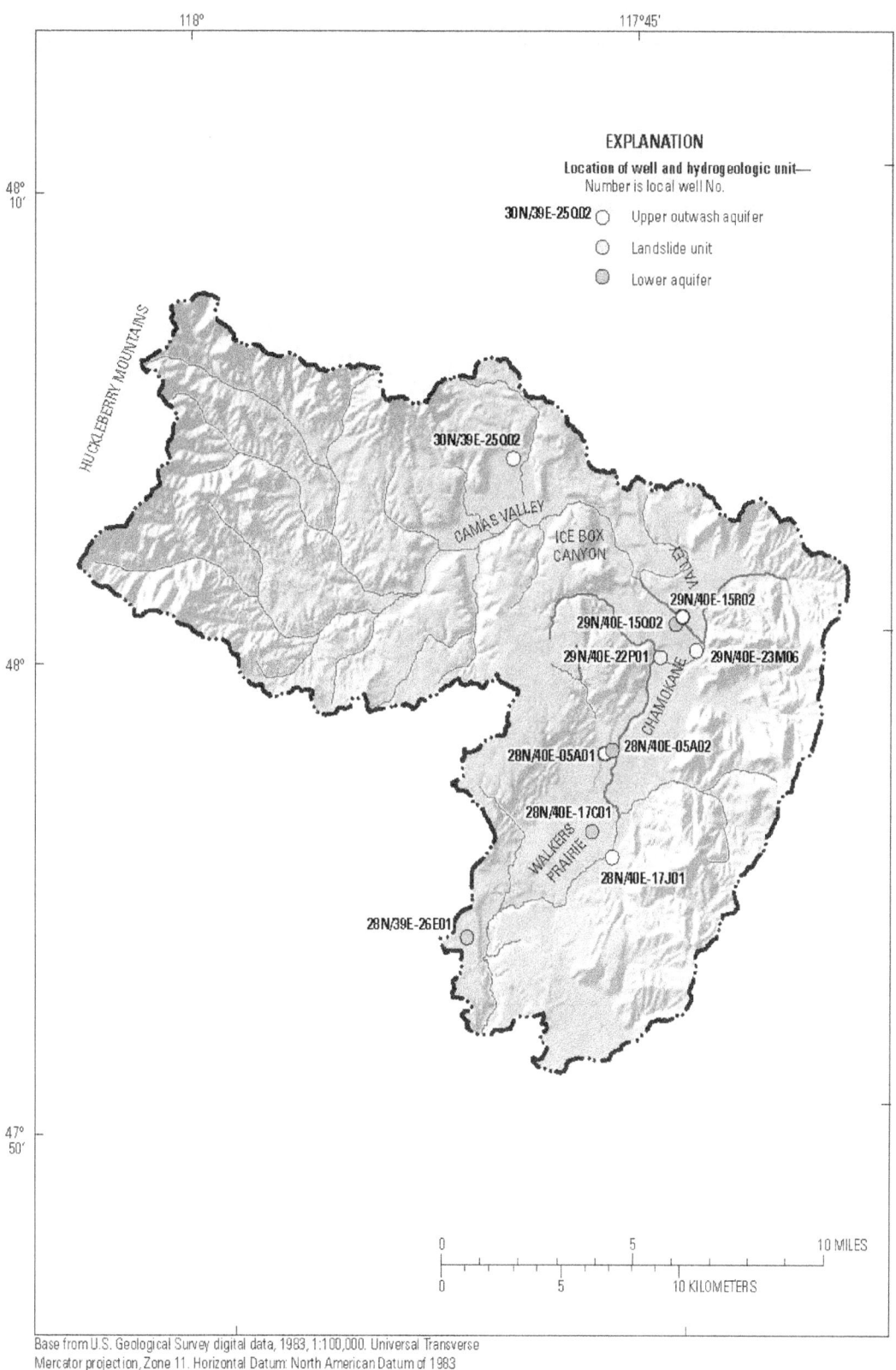

Figure 19. Location of selected wells with measured and simulated water-level altitudes, Chamokane Creek basin, Washington.

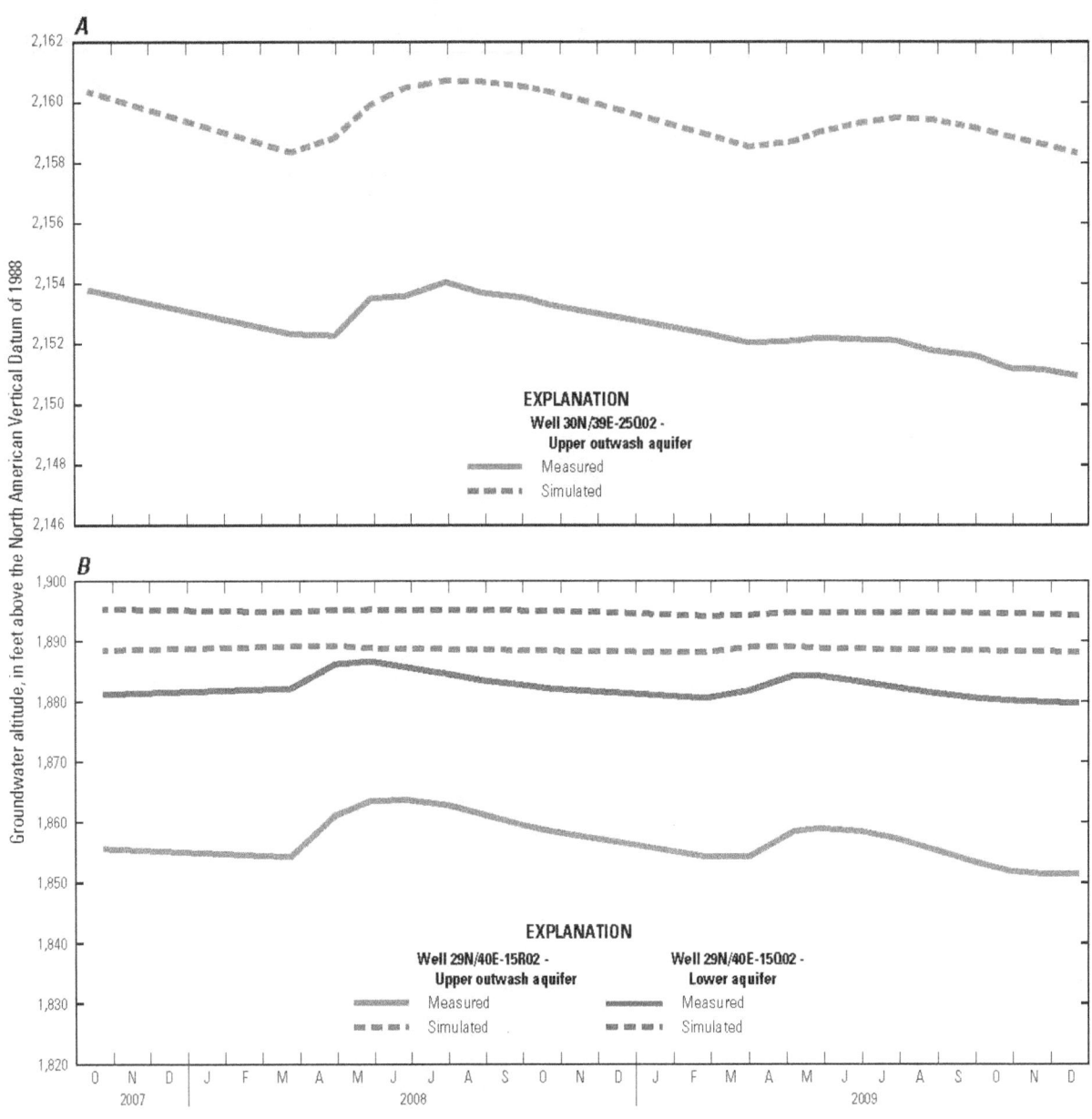

Figure 20.   Measured and simulated water-level altitudes in selected wells, Chamokane Creek basin, Washington.

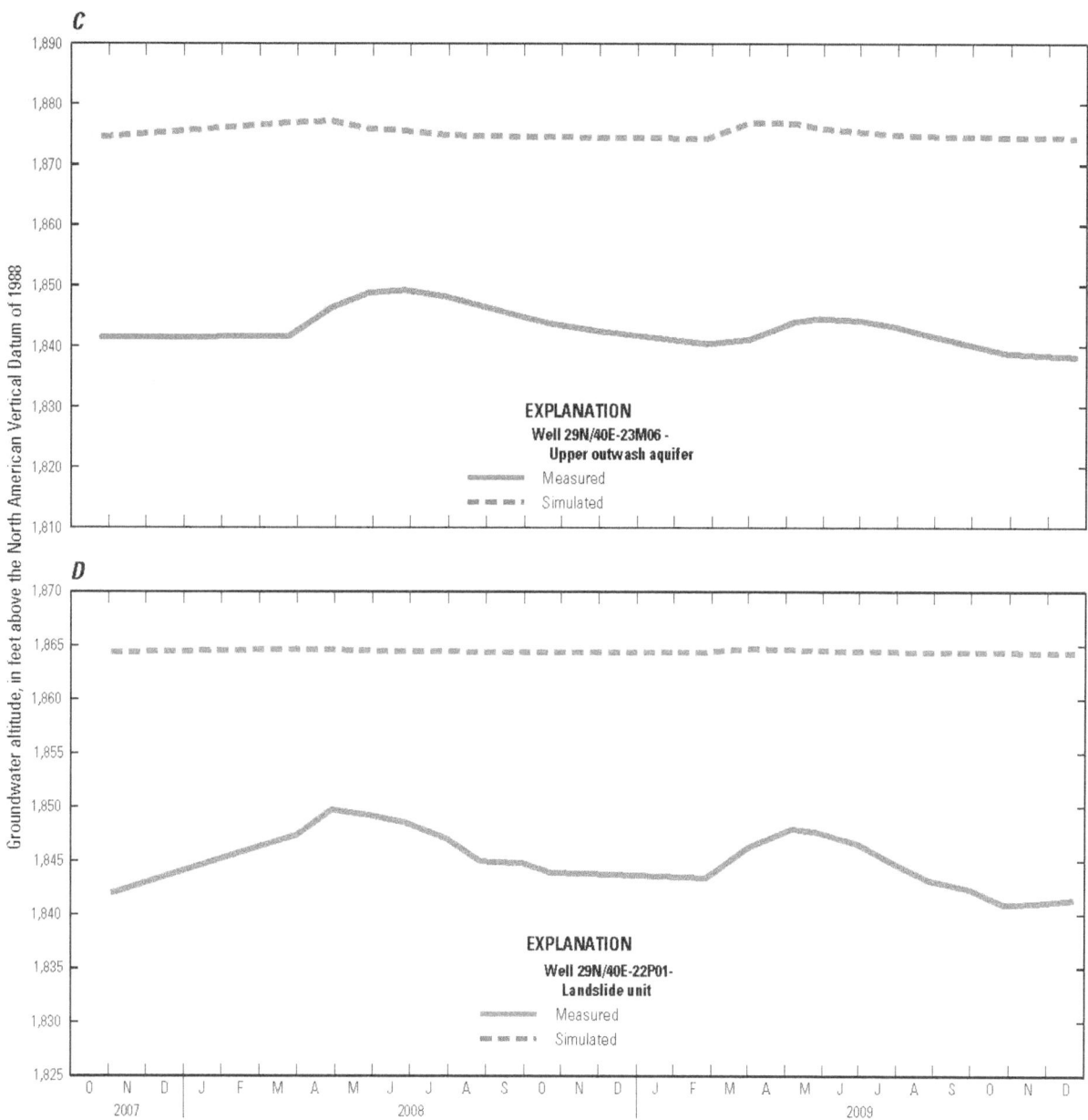

Figure 20.—Continued

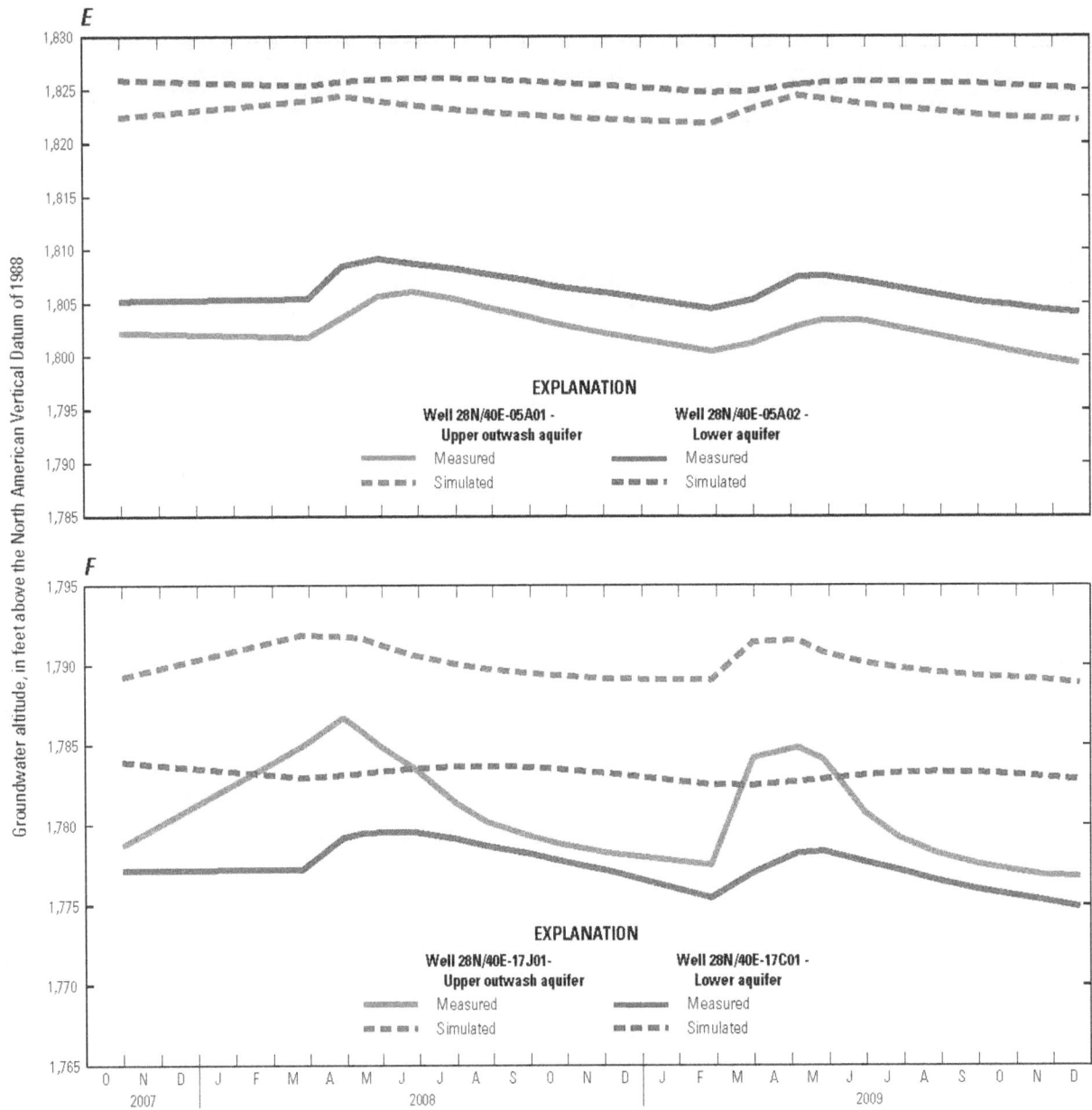

Figure 20.—Continued

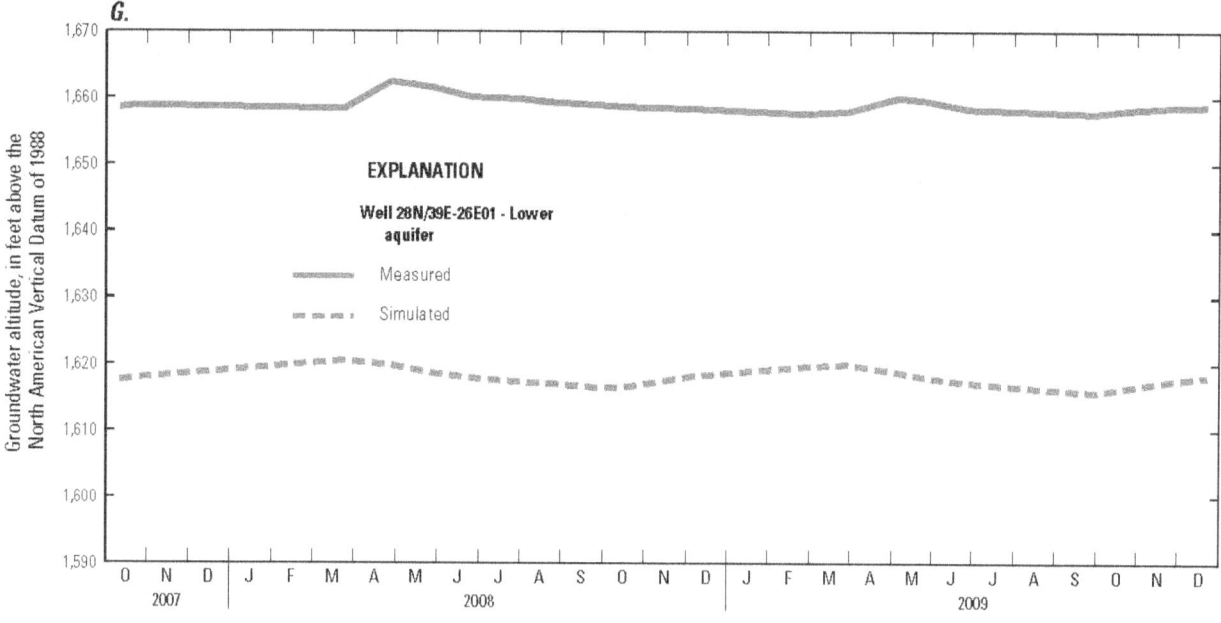

Figure 20.—Continued

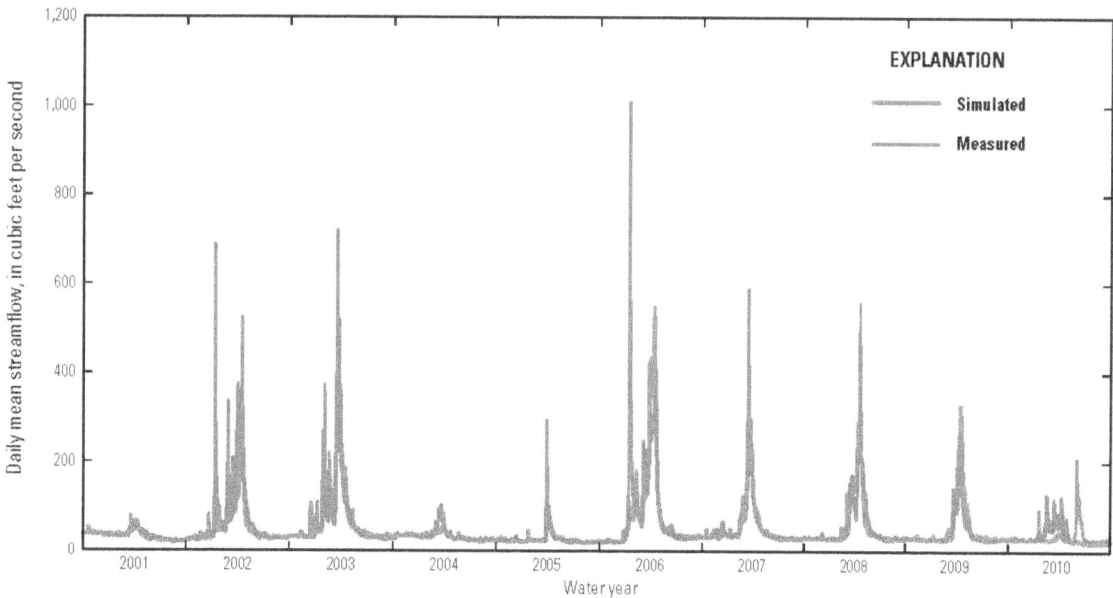

Figure 21. Simulated and measured daily mean streamflow for the streamflow-gaging station in the Chamokane Creek basin, Washington, water years 2000–2010.

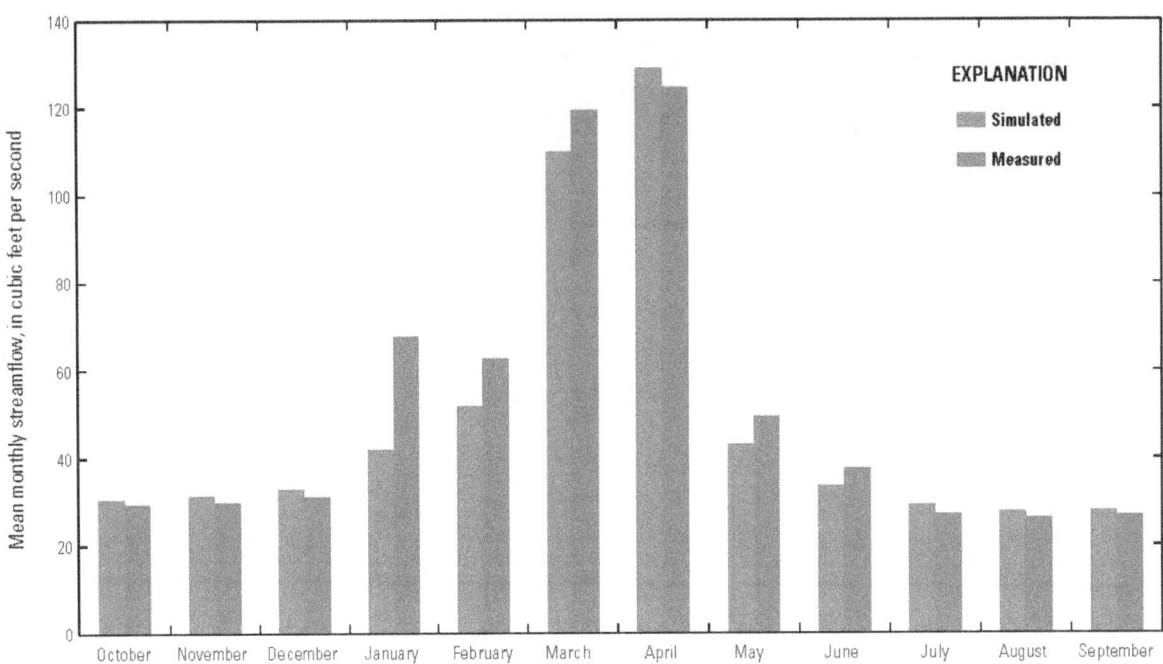

Figure 22. Simulated and measured mean monthly streamflow for streamflow-gaging station in the Chamokane Creek basin, Washington, water years 2000–2010.

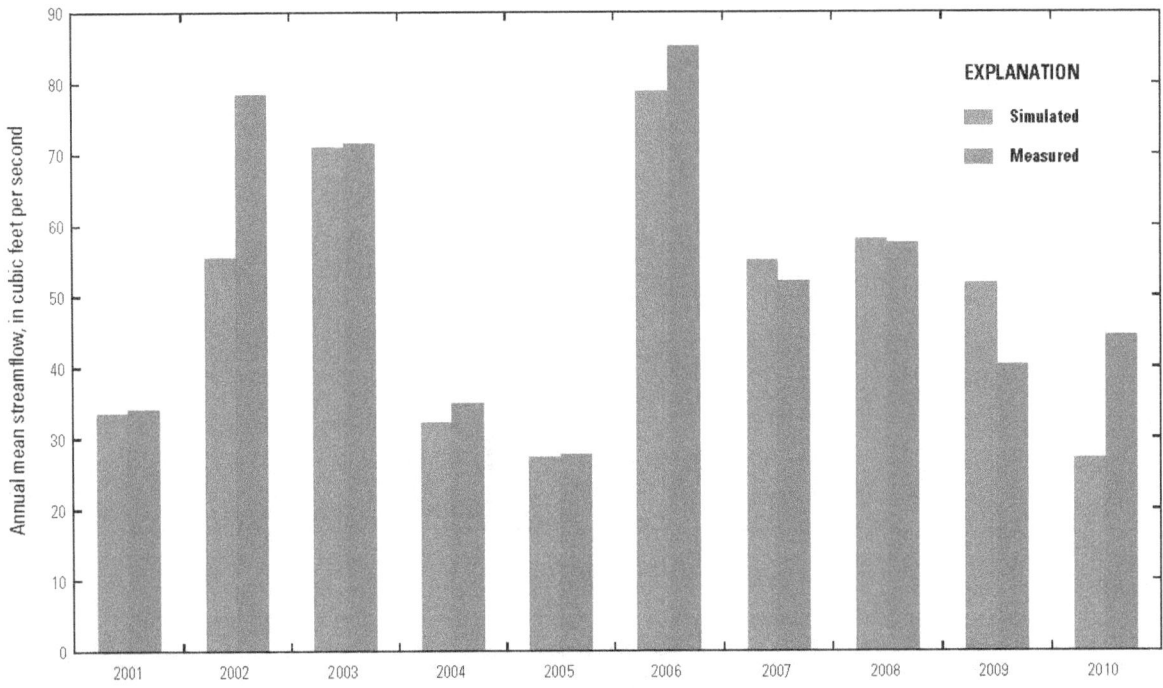

Figure 23. Simulated and measured annual mean streamflow for the streamflow-gaging station in the Chamokane Creek basin, Washington, water years 2000–2010.

Table 6. Volumetric efficiency by water year and average volumetric efficiency by month, Chamokane Creek basin, Washington.

Water year	Volumetric efficiency	Month	Volumetric efficiency
1999	0.74	January	0.76
2000	0.69	February	0.74
2001	0.89	March	0.65
2002	0.66	April	0.72
2003	0.74	May	0.78
2004	0.86	June	0.79
2005	0.86	July	0.83
2006	0.65	August	0.85
2007	0.82	September	0.85
2008	0.75	October	0.89
2009	0.69	November	0.88
2010	0.59	December	0.85

Model Uncertainty and Limitations

The Chamokane Creek model represents a complex natural system perturbed by human activities, with a set of mathematical equations that describe the system. Intrinsic to the model is the error and uncertainty associated with the approximations, assumptions, and simplifications that must be made. In addition to those intrinsic errors, hydrologic modeling errors typically are the consequence of a combination of errors in the (1) input data, (2) representation of the physical processes by the algorithms of the model, and (3) parameter estimation during the calibration procedure (Troutman, 1985). These three types of model errors within the model and how those errors limit application of the model are as follows:

1. Data on types and thicknesses of mapped hydrogeologic units, pumpage, hydraulic properties, land use/land cover, forest density, and soil types were taken from many sources at different levels of refinement and for various intended purposes. Most of the measured data were concentrated along the Chamokane Creek valley and populated areas because that is where most wells are located. This means that for some of the study area, information is unavailable to constrain the model, especially for the areas lacking water-level data.

    Portions of the model domain include basin-fill sediments, basalt, and bedrock, which are unmapped or poorly characterized. In areas without lithologic well logs, variability in hydrogeologic properties or depths of contacts may fall outside the range of values in areas that

have been better characterized, and the errors associated with this variability would remain unrepresented. Specific conclusions drawn from regions of the model with sparse observations should be limited to general flow directions and relative magnitudes.

The initial hydraulic-property data generally came from specific-capacity tests, which typically measure drawdown at one time and at one pumping rate, and are not as accurate as aquifer tests. Thus, broad ranges of hydraulic-property parameter values are possible, especially for the deeper part of the flow system. Lack of information on streambed hydraulic conductivity values resulted in these values being poorly constrained, which may limit the accuracy of groundwater/surface-water exchanges.

The PRMS watershed model requires measured precipitation and air temperature time-series data and physical characteristics of the basin. Precipitation volume is often the most important driving factor of the simulation, and it is often the most difficult to estimate. Precipitation records are point measurements, whereas the model requires input distributed throughout the study area. This study used four precipitation sites within and adjacent to the study area, and the measurements were extrapolated to estimate precipitation throughout the entire basin. Precipitation in the Chamokane Creek basin varies widely. Mean altitude of an HRU can differ significantly from that of the closest rain gage, and the HRU can include a wide range of average precipitation. In addition to the problems with spatial distribution, much of the precipitation comes in the form of snowfall, which can be underestimated if the collection device is not protected from the wind. Catchment losses also occur for rain, but they are believed to be smaller than for snow. Temperature data can be the source of as much potential error as the rainfall data. Again, temperature is recorded as a point measurement and basin-wide distributed values must be estimated for each HRU. Differences of a few degrees can determine if precipitation is simulated as snow or rain or if an accumulated snowpack melts. Precipitation, combined with air temperature, determines both the cumulative annual streamflow and the basic shape of the simulated hydrograph.

In general, the DEM and the GIS Weasel represented the physical characteristics of the basin well. Even though the basin was delineated into HRUs, approximations of slope and aspect were necessary. Coarse coverages of forest density, land use, and soils introduced error in sensitive parameters that determine ET, infiltration, and groundwater recharge.

2.   A numerical model can not completely represent all physical processes within a drainage basin. Determining if a weakness in a simulation is attributable simply to input data error or shortcomings in how the model represents the governing physical processes is intractable. The model inevitably relies on simplifying assumptions and generalizations that complexly affect the results of the simulation. The Chamokane Creek model was not designed to represent every detail of the hydrologic system, and simulation results will vary based on which details were and were not emphasized. For example, small differences in simulated heads in the upland areas can result in large differences in simulated tributary streamflow because the complex nature of the stream system and valleys in these areas was not represented in the model.

Model-discretization errors result from (1) the effects of averaging altitude information over the model cell size, (2) the time-averaging of modeled stresses inherent in a 6-month simulation stress period (although two of the most climatic factors—precipitation and air temperature—are modeled using daily time steps), and (3) the inaccuracies in the geometric representation of mapped hydrogeology. For example, the modeled land-surface altitude was derived from the resampled 10-meter DEM to create one altitude for the uppermost model cell. However, the ranges in actual altitudes across a model cell were as much as 675 ft and averaged 165 ft. For this reason, interpretations of simulation results should be limited to scales several times greater than the model spatial and temporal resolutions of 1,000 ft and 1 day, respectively, or larger in areas of steep surface gradients or rapidly changing conditions.

3.   Errors in parameter estimates occur when improper values are selected during the calibration process. Various combinations of parameter values can result in low residual error, yet improperly represent the actual system. An acceptable degree of agreement between simulated and measured values does not guarantee that the estimated model parameter values uniquely and reasonably represent the actual parameter values. The use of automatic parameter estimation techniques and associated statistics, such as composite scaled sensitivities and correlation coefficients, removes some of the effects of non-uniqueness, but certainly does not eliminate the problem entirely. The comparison of calibrated values to literature values also can reduce error caused by parameter estimation if the model results are within previously accepted ranges. Limitations of the observation weighting scheme used in this study include non-varying weights for heads and seasonal streamflow that did not take into account measurement errors within each group of measurements.

Model-Derived Hydrologic Budgets

The Chamokane Creek basin coupled groundwater- and surface-water flow model can be used to derive components of the hydrologic budget for the simulation period (1980–2010). During this period, the distribution and amount of pumpage changed and climate varied, and a cumulative or mean annual water budget would not highlight these variations. Although the model calculates all budget components at the daily time step, short-term variability also can mask general trends. Thus, simulated water budgets are presented as annual values (fig. 24).

The simulated water budgets show the variations between the range of climatic conditions, including extremely dry years (1994 and 2001) and extremely wet years (1983 and 1997). During dry years, almost all available water is evapotranspired, leaving little to recharge the groundwater system. During wet years, available precipitation greatly exceeds actual evapotranspiration, and groundwater recharge and streamflow increase.

Average annual precipitation estimated by the model was 24.3 in., which is more than the PRISM-derived average precipitation of 19.6 in. (Kahle and others, 2010). Simulated precipitation in the high-altitude areas was adjusted upward to produce the spring snowmelt-driven discharge and these parameter adjustments also resulted in higher precipitation in the lower altitude areas along the valley floor (fig. 25). Simulated actual evapotranspiration was 79 percent of precipitation, which compared well to the value reported in Kahle and others (2010; 76 percent). At times, streamflow exceeds the difference between precipitation and ET due to hatchery return flows. Hatchery operations were simulated by specifying surface-water return flows at hatchery locations equal to the groundwater pumping rate. Mean annual groundwater recharge was 4.8 in/yr.

The Chamokane Creek model also can be used to derive components of the groundwater budget for each of the stress periods for the 31-year simulation. Simulated water budgets are presented for a wet (2006), average (2008), and dry (2005) year. These three years capture the hydrologic variability present in the model domain and are representative of the existing conditions. As measured at USGS streamflow-gaging station, Chamokane Creek below Falls, near Long Lake (12433200; the most downstream streamflow site in the basin), the ratios of the annual mean discharge (sum of the daily mean discharges for one year divided by the number of days in that year) to the mean annual discharge (sum of the annual mean discharges divided by the number of years; calibration period of 1988–2010) for 2006, 2008, and 2005 were 1.45, 0.98, and 0.47, respectively. These ratios show that these years are representative of a wide range in climatic conditions, and thus also a wide range in recharge.

The simulated water budgets (fig. 26) show the variations in inflows and outflows between the three types of climatic years. Inflows are flows of water into the aquifer system (fig. 26A). Outflows are flows of water out of the aquifer system (fig. 26B). For example, recharge is considered an inflow and pumping wells are considered an outflow. Groundwater recharge ranged from about 29 ft³/s in 2005 to 105 ft³/s in 2006, a 262-percent increase compared to a dry year. In 2005, less water flowed out of the groundwater system into storage because water levels were lower in the dry year (2005) than in the wet (2006) and average (2008) years. Leakage to streams and to the land surface was greater during the wet year and this was the source of the higher annual streamflow, as groundwater discharged to the surface was available to be routed overland to streams. The difference between the inflow and outflow to streams is least during the wet year in 2006 (fig. 26C). Although this is counterintuitive, it is due in some part to the dry antecedent conditions following 2005 (dry year). Stream stages also are higher during the wet years and if water levels are slower to respond to the wet conditions, differences between stream stage and groundwater levels would be greatest, and thus produce conditions for greater streamflow loss.

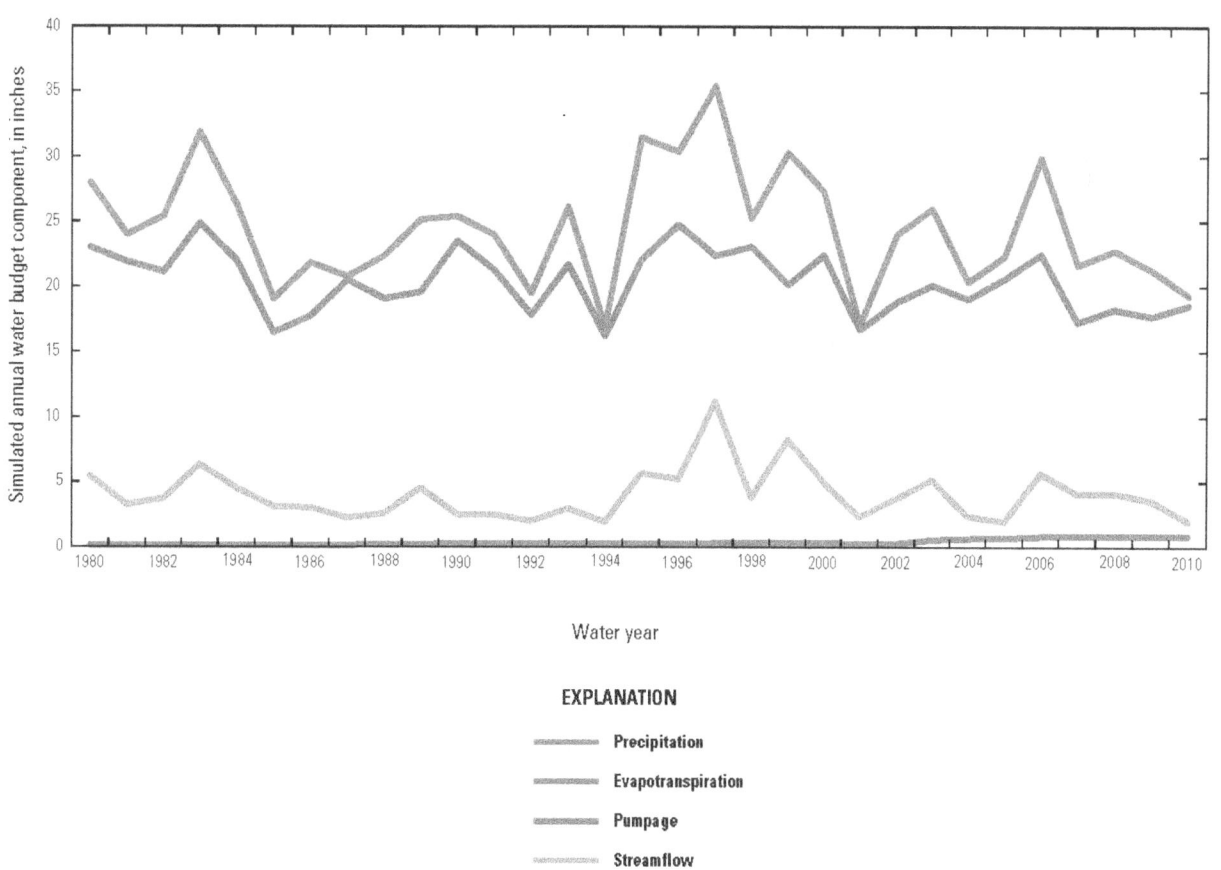

Figure 24. Simulated annual water budgets, Chamokane Creek basin, Washington, 1980–2010.

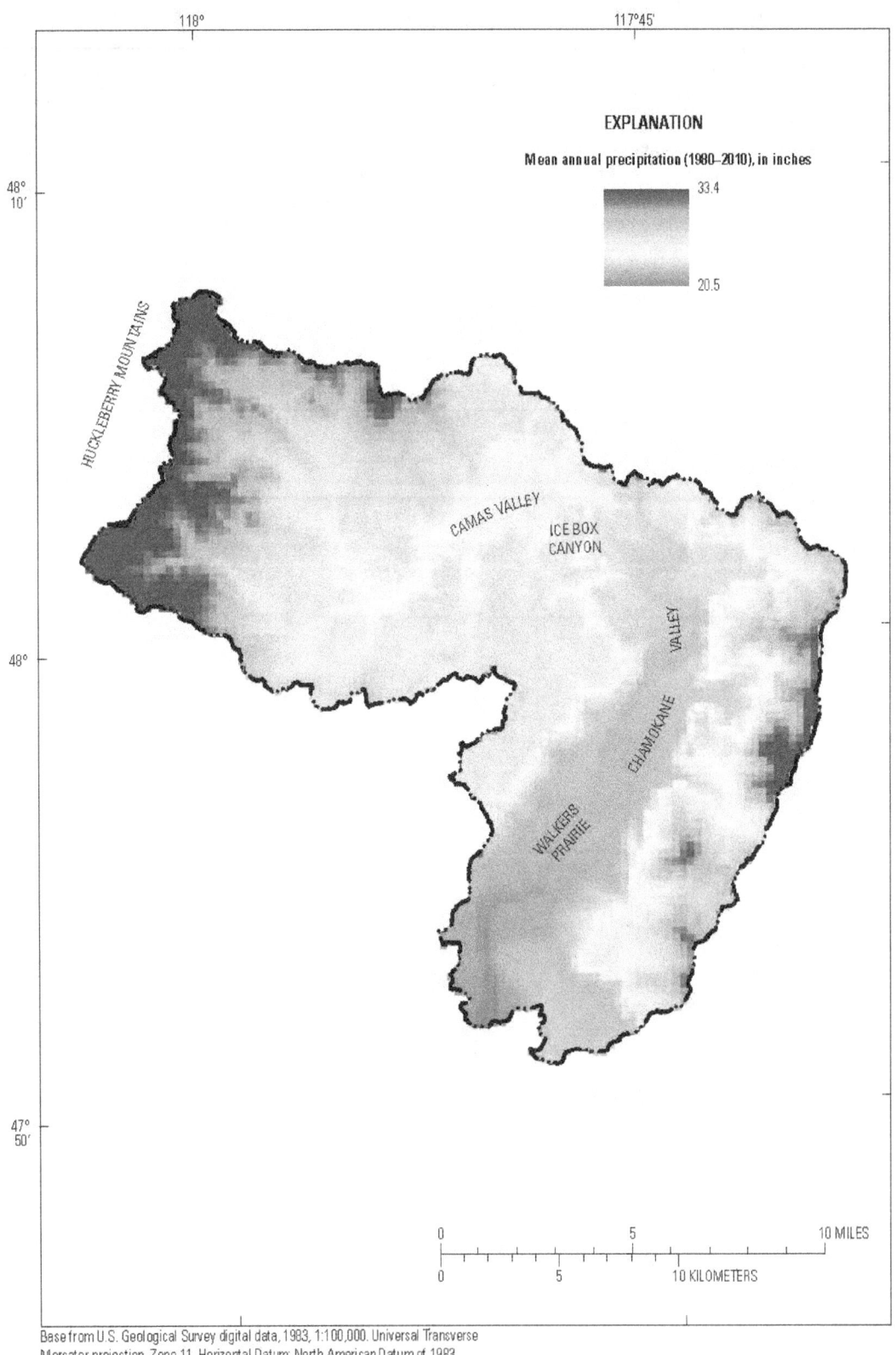

Figure 25.   GSFLOW-derived mean annual precipitation (1980–2010), Chamokane Creek basin, Washington.

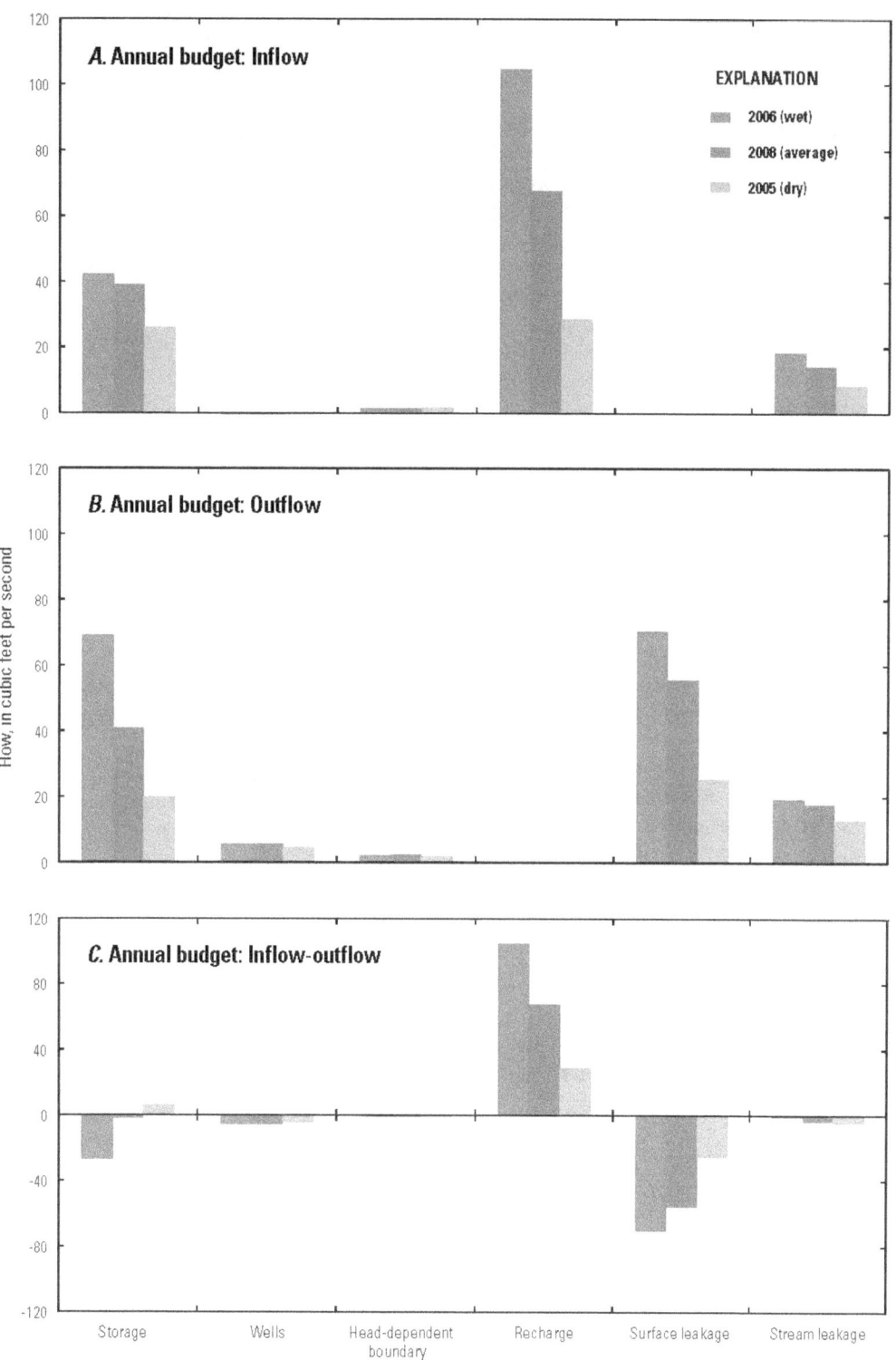

Figure 26.   Simulated annual water budgets for wet (2006), average (2008), and dry (2005) years, Chamokane Creek basin, Washington.

Evaluation of Water-Management Alternatives

The Chamokane Creek model was used to estimate the response of the regional flow system to potential changes in stresses. These management alternatives, or scenarios, are used to better understand the relation of the groundwater system to surface-water resources. In particular, the scenarios simulate the relation between pumping of groundwater and streamflow. The potential effects of scenarios are assessed by comparing simulated output from the scenarios with simulated output from the calibrated model; that is, streamflow simulated from a scenario is compared to the base case simulated streamflow. All assumptions and limitations underlying the base case model are assumed to apply for the scenarios as well.

The scenarios are used to better understand the relation between groundwater pumping and Chamokane Creek streamflow during water years 1999–2010 and address the specific factual questions filed by the U.S. District Court. Comparison of the calibrated model (base case) results to scenario results provide the framework for assessing the potential effects; that is, streamflow simulated in a scenario can be compared to the simulated base case streamflow along the simulated stream network. This framework analyzes changes in streamflow and is well-suited to the intended use of the model.

The factual questions reference the upper, middle, and lower Chamokane Creek basins. However, only the areas referred to as upper and middle Chamokane Creek basin are actually within the model domain, the outlet of which is located at the granite dike that forms Chamokane Falls (United States v. Anderson, 1979). The outlet of the basin also is marked by the USGS streamflow-gaging station 12433200, located immediately downstream of the Chamokane Falls. This gage has been the point of regulation since the adjudication of the basin. The small area between the falls and the mouth of the creek at the Spokane River is known as the Lower Chamokane area (United States v. Anderson, 1979). The lower basin area contains sediments/aquifers that are largely unsaturated and become intermixed with a different aquifer system. The dividing point for upper and middle basins is Ice Box Canyon (fig. 27).

For these reasons, effects of upper and middle basin water use at the USGS streamflow-gaging station are documented here, but water use in the lower basin downstream of the gage was neither simulated nor documented. Regulation of junior rights occurs for streamflow at the USGS streamflow-gaging station at the basin outlet, so all scenario results are the difference in simulated streamflow at this location.

The scenario simulation period (1999–2010) allows for a temporal assessment that accounts for changes in pumping over time. Explicitly included in the model is a large range in climatic conditions and thus, streamflow and natural recharge. As a result, a broad range of hydrologic conditions (both natural and human induced) are simulated in the model, which in turn is represented in the simulated streamflow for Chamokane Creek.

Model scenarios were simulated using the calibrated model to address the factual questions of the court (excerpted from United States v. Anderson, 2006).

1. **Factual Question**: Is the groundwater of the upper basin separate or connected from that of the middle and lower basin areas?

2. **Factual Question**: Do all surface-water and groundwater uses in the middle and lower Chamokane areas impact flows in Chamokane Creek?

3. **Factual Question**: What are the cumulative impacts of claims registry use and permit-exempt wells on the flow in Chamokane Creek?

4. **Factual Question**: If there are any impacts identified in questions 2 and 3 that are sufficiently large to affect the flows, how do those impacts affect the frequency and severity of regulation by the Water Master?

4. **Factual Question**: Is there a level of domestic or stockwater use that is too small or difficult to regulate? If so, what is that level?

Scenario 1 – Connection of the Upper and Middle Basin Groundwater Flow Systems

1. **Factual Question**: Is the groundwater of the upper basin separate or connected from that of the middle and lower basin areas?

Approach: This scenario is formulated to better understand the relation between groundwater pumping in the upper basin model domain and surface-water resources. All upper basin groundwater withdrawals and associated septic returns were increased by 100 percent to determine current connection within the middle basin. The difference in streamflow and groundwater levels between the base case and scenario results is the simulated effect of upper basin withdrawals on the middle basin.

Streamflow for the 12-year period (1999–2010) was simulated using the Chamokane Creek model operated with twice the upper basin groundwater pumping. Upper basin mean annual pumping during the 12-year scenario period was increased about 0.08 ft³/s (58 acre-ft; table 7). All other model stresses remained the same as the calibrated model. Public water supply was the largest category of water use.

The difference in simulated streamflow at the USGS streamflow-gaging station is an indicator of the effects of upper basin withdrawals on the middle basin. Mean annual difference in streamflow between the calibrated model and scenario 1 streamflows was a decrease in flow of 0.05 ft³/s. This simulated difference in streamflow is small and within expected model error, but represents a large percentage of the total mean annual pumping and surface-water diversions. Monthly mean differences in streamflow also are small and show variability due to timing of effects, therefore it is difficult to see trends.

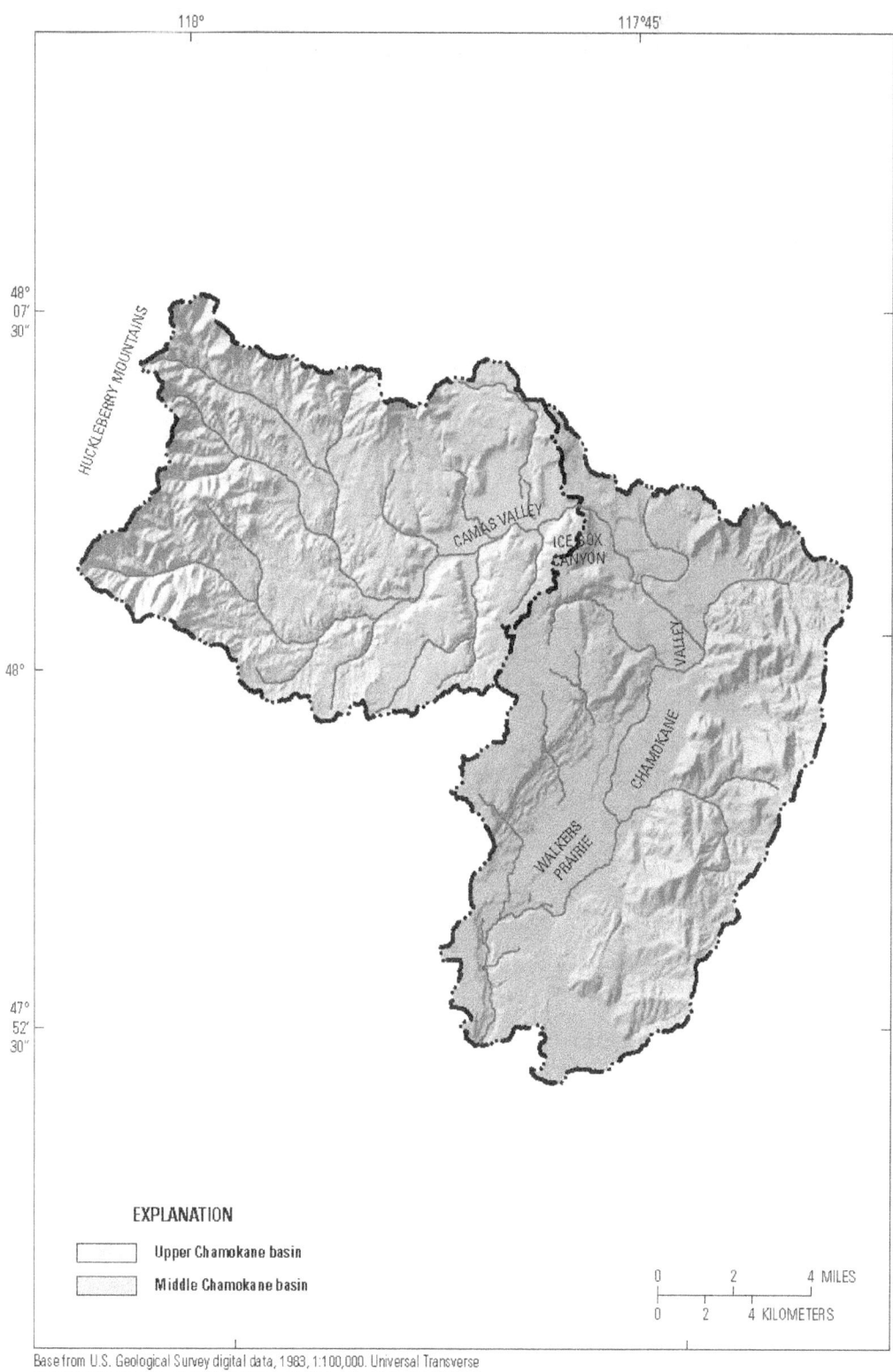

Figure 27.   Location of the upper and middle Chamokane Creek basins, Washington.

To better demonstrate the effects of upper basin pumping on streamflow, the cumulative monthly mean difference (monthly mean difference added to previous monthly mean difference) in streamflow is shown in figure 28.

On a monthly basis, the largest effects generally occur from January through April, but the effects of water use are fairly equally distributed over time. Effects of groundwater pumping are attenuated over time depending on the distance from a simulated surface-water feature. The scenario results show periods of streamflow increases (positive differences) due to the increases in simulated septic return flows. The model simulates 52 percent of permit-exempt and claims registry use as non-consumptive (Vaccaro and Olsen, 2007; Ely and others, 2011). Groundwater is withdrawn from deeper model layers and returned to the land surface as septic-system return flow, where it becomes available to streamflow. Increasing these septic returns can cause increases in streamflow, but as the mean annual difference shows, long-term effects of the doubling of upper basin pumping is a net decrease in streamflow. This effect will be seen in other scenario results.

Average simulated water-level change in the Upper outwash aquifer was about 0.1 ft in the upper basin and slightly less in the middle basin. These results suggest there is a connection between the groundwater systems of the upper and middle basins, but the effects of 2010 groundwater withdrawals in the upper basin are small due to the small stress on the system. Connection between the groundwater systems of the upper and middle basins also can be seen with an examination of water levels. Figure 5 shows inferred directions of groundwater flow from the water-level contours that suggest groundwater movement from the upper to middle basins in the Upper outwash aquifer.

Table 7. Simulated net mean annual groundwater pumping; surface-water diversions and returns; and mean annual changes in net groundwater pumping, surface-water diversions and returns, and streamflow from base case for model scenarios, 1999–2010, Chamokane Creek basin, Washington.

[Net groundwater pumping values represent groundwater pumpage minus non-consumptive return flows. Negative values represent a decrease from base case. Positive values represent an increase from base case. **Abbreviation:** ft^3/s, cubic feet per second]

Scenario No.	Net groundwater pumping (ft^3/s)	Surface water		Change in			
		Diversions (ft^3/s)	Returns (ft^3/s)	Net groundwater pumping (ft^3/s)	Surface-water diversions (ft^3/s)	Surface-water returns (ft^3/s)	Streamflow (ft^3/s)
Base case - Calibrated model at existing conditions	4.04	0.02	3.08				
Scenario 1 - 100 percent increase in upper basin groundwater pumping and associated septic returns	4.12	0.02	3.08	0.08	0.00	0.00	−0.05
Scenario 2A - No middle basin groundwater or surface-water withdrawals	0.02	0.01	0.00	−4.02	−0.01	−3.08	0.28
Scenario 2B - No middle basin groundwater and surface-water withdrawals except existing hatchery operations	3.10	0.01	3.08	−0.94	−0.01	0.00	0.79
Scenario 3 - No claims registry or permit exempt water use	4.02	0.02	3.08	−0.02	−0.002	0.00	0.02
Scenario 4A - No groundwater pumping and surface-water diversions for all water-use categories except hatchery operations	3.08	0.00	3.08	−0.96	−0.02	0.00	0.81
Scenario 4B -100 percent increase in groundwater pumping and surface-water diversions for all water-use categories except existing hatchery operations	5.00	0.04	3.08	0.96	0.02	0.00	−0.81
Scenario 5A - No permit-exempt pumping and stockwatering from groundwater pumping and surface-water diversions	4.02	0.02	3.08	−0.02	0.00	0.00	0.02
Scenario 5B - No stockwatering	4.03	0.00	3.08	-0.01	−0.02	0.00	0.02

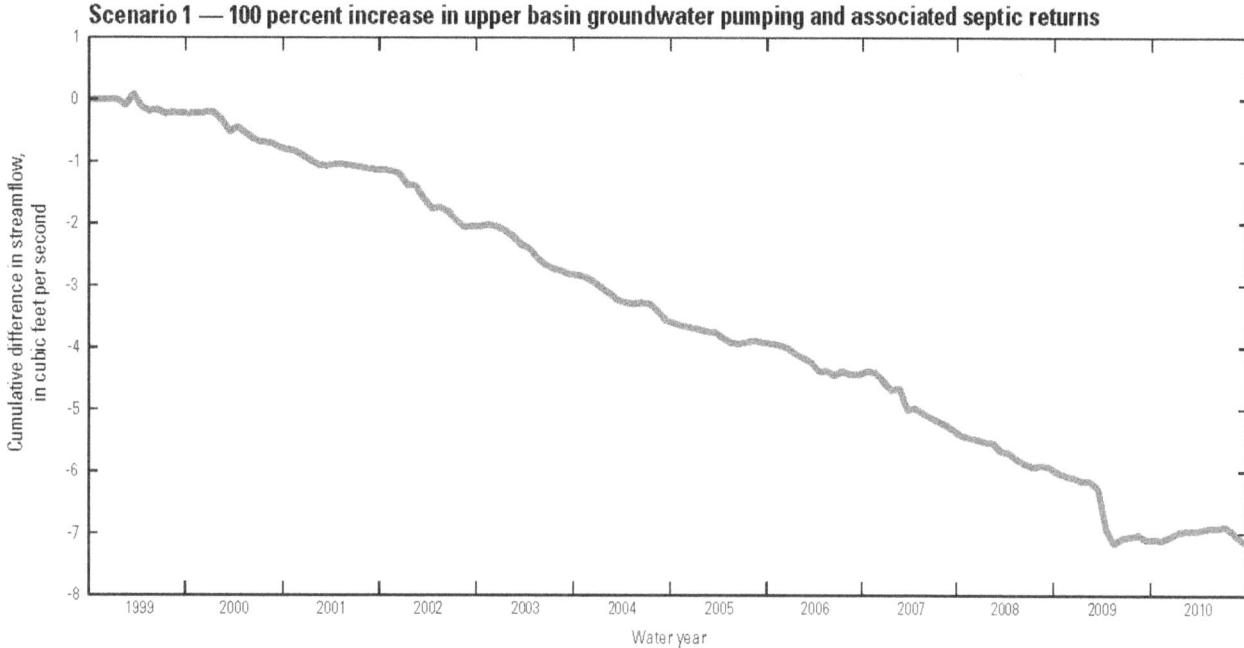

Figure 28. Cumulative difference in simulated monthly mean streamflow between existing conditions and existing conditions with increased upper basin groundwater pumping, Chamokane Creek basin, Washington.

Scenario 2 – Effect of Surface and Groundwater Uses in the Middle Basin on Chamokane Creek Streamflow

2. **Factual Question**: Do all surface-water and groundwater uses in the middle and lower Chamokane areas impact flows in Chamokane Creek?

    Approach: This scenario is formulated to better understand the relation between current water use in the middle basin and Chamokane Creek streamflow. First, all groundwater pumping and surface-water withdrawals and returns in the middle basin were turned off in the model and the difference in streamflow between the base case and scenario results is the simulated effect of middle basin water use on streamflow. Second, all groundwater pumping and surface-water withdrawals, except those associated with hatchery operations, were turned off in the model. Surface-water and groundwater uses in the lower basin were outside the model domain and were not simulated.

    Streamflow for the 12-year period (water years 1999–2010) was simulated using the Chamokane Creek model operated without middle basin groundwater or surface-water withdrawals. Middle basin mean annual pumping during the 12-year scenario period that was eliminated was about 4.04 ft³/s (2,896 acre-ft; table 7), and ranged from a minimum of about 2 ft³/s (1,448 acre-ft) in 1999 to a maximum of

about 7 ft³/s (5,068 acre-ft) in 2008. Mean annual middle basin surface-water diversions and hatchery returns that were eliminated were 0.01 ft³/s (7 acre-ft; table 7) and 3.08 ft³/s (2,230 acre-ft; table 7), respectively. The largest category of water use in the middle basin was groundwater pumping for hatchery operations, followed by public water supply.

The difference in simulated streamflow at the USGS streamflow-gaging station is an indicator of the effects of middle basin water use on the Chamokane Creek streamflow. The monthly mean difference in streamflow is shown in figure 29A. The mean annual difference in streamflow between the calibrated model and scenario 2A streamflows was 0.28 ft³/s (table 7). Effects of surface-water diversions are seen almost immediately, whereas effects of groundwater pumping are attenuated over time depending on the distance from simulated surface-water features. Presumably, streamflow would continue to increase with time until the near-full effect of turning off the wells (that is, 4.04 ft³/s) was realized in the stream.

Results of this scenario are driven by the overwhelming effect of hatchery operations. The two hatcheries in the middle basin account for most of the groundwater pumping but also augment streamflow by returning the groundwater directly to streams. Unlike the effects of surface-water return flow, the effects of the groundwater pumping are not seen immediately in the simulated streamflow. The summertime increase in hatchery pumping and of surface-water return flow produces the alternating pattern of streamflow increases (positive) and decreases (negative) shown in figure 29A.

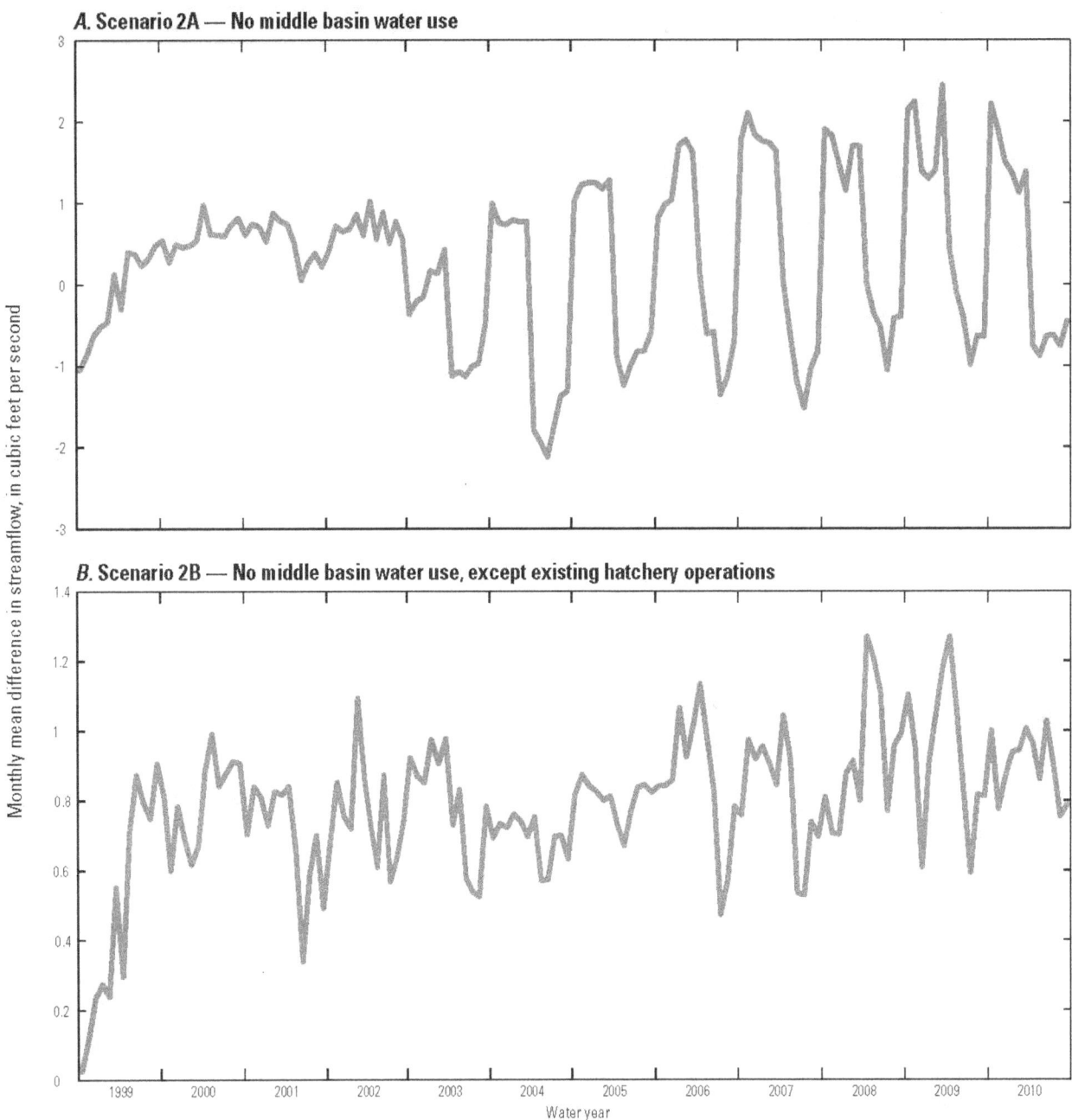

Figure 29.   Difference in simulated monthly mean streamflow between existing conditions and existing conditions without middle basin water use and without middle basin water use except existing hatchery operations, Chamokane Creek basin, Washington.

To separate the effects of hatchery operations, scenario 2 also was run with no middle basin pumping or diversions, except for the pumping and return flow associated with hatchery operations. The monthly mean difference in streamflow is shown in figure 29B. The mean annual difference in streamflow between the calibrated model and scenario 2B with normal hatchery operations was 0.79 ft³/s (table 7), almost equal to the average annual difference in pumping between existing and scenario 2B conditions (0.9 ft³/s; table 7). Based on these results, it is apparent that surface and groundwater uses in the middle basin impact flows in Chamokane Creek.

Scenario 3 – Cumulative Impacts of Claims Registry and Permit-Exempt Wells on Chamokane Creek Streamflow

3. **Factual Question**: What are the cumulative impacts of claims registry use and permit-exempt wells on the flow in Chamokane Creek?

   Approach: This scenario was approached in a similar manner as the scenarios in response to Factual Question 2. All groundwater and surface-water withdrawals designated as "claims registry use" and "permit-exempt" were turned off in the model and the difference in streamflow between the base case and scenario results is the simulated effect of claims registry and permit-exempt withdrawals on streamflow.

   The calibrated model was used to simulate the 12-year period (1999–2010) without claims registry and permit-exempt water use. Claims and permit-exempt usage is mostly self-supplied domestic withdrawals with some stock watering and small-scale irrigation. The total mean annual claims registry and permit-exempt use was 0.03 ft³/s (22 acre-ft; table 7) and included both groundwater and surface-water withdrawals. For this scenario, the calibrated model was run and the resulting simulated streamflows were then used as the base case for assessing the relation between claims registry and exempt pumpage with surface-water resources. Septic return was not simulated when the claims and permit-exempt pumpage was not simulated. The cumulative difference in monthly mean streamflow between existing conditions and conditions without claims registry use and permit-exempt pumping is shown in figure 30. Mean annual difference in streamflow was 0.02 ft³/s (table 7). This value represents the cumulative impacts of claims registry use and permit-exempt wells on flow in Chamokane Creek. Based on these results, it is apparent the impact of claims registry and permit-exempt wells on Chamokane Creek is proportional to the rate of groundwater pumping.

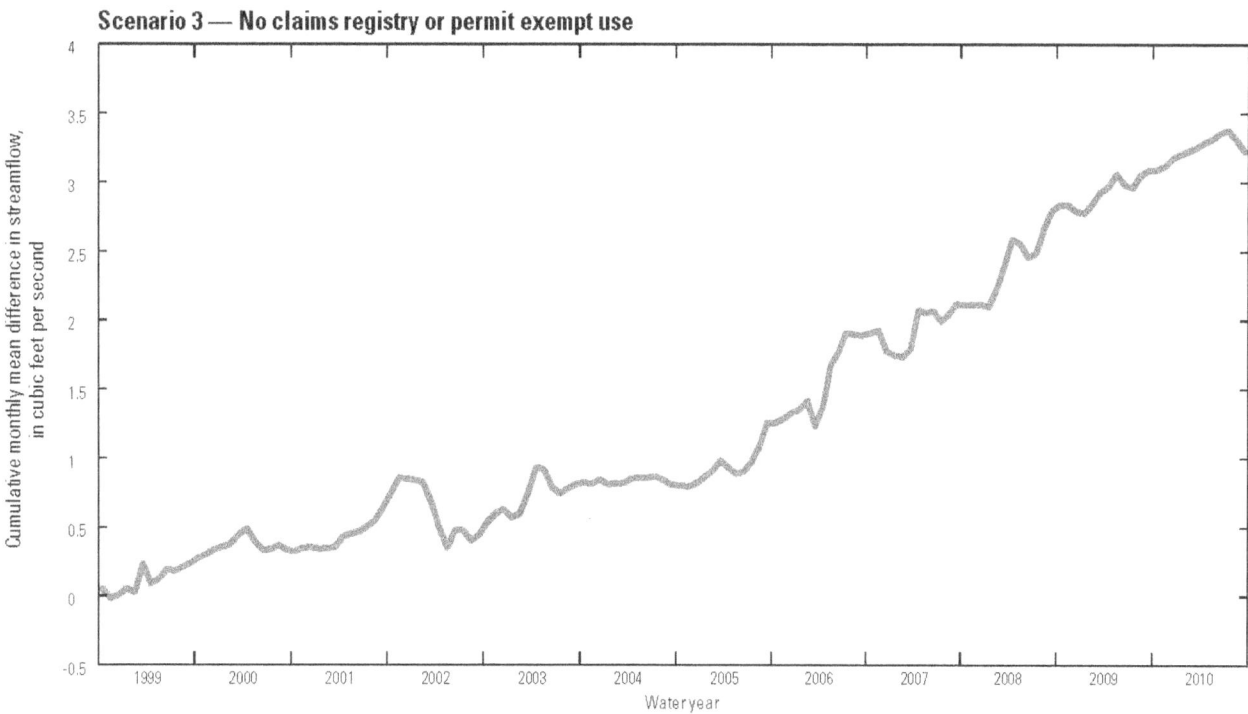

Figure 30. Cumulative difference in simulated monthly mean streamflow between existing conditions and existing conditions without claims groundwater pumping and surface-water diversions and returns and permit-exempt pumping, Chamokane Creek basin, Washington.

Scenario 4 – Frequency of Regulation Due to Impacted Streamflow

4.   **Factual Question**: If there are any impacts identified in questions 2 and 3 that are sufficiently large to affect the flows, how do those impacts affect the frequency and severity of regulation by the Water Master?

    Approach: If it is determined that streamflow is affected by groundwater pumping, the simulated change in streamflow would be compared to measured streamflow at the USGS streamflow-gaging station. Regulation by the Water Master occurs when the 7-day low flow is less than 24 ft³/s. Historical streamflow records for Chamokane Creek (station 12433200) were examined to determine how often streamflow has been less than 24 ft³/s in the past, and then compared to the historical frequency of occurrence with simulated streamflows that include impacts to streamflow from groundwater pumping.

    Groundwater pumping and surface-water diversions for all water-use categories except hatcheries were first eliminated and then doubled and associated changes in streamflow were determined. The approach demonstrated the connection between water use and streamflow at different withdrawal rates. The court-appointed Water Master regulates junior water rights when the mean daily 7-day low flow becomes less than 24 ft³/s (27 ft³/s for rights issued after December 1988) at Chamokane Falls, as recorded at USGS streamflow-gaging station 12433200. Figure 31 shows the six periods from water years 1999–2010 when the mean daily 7-day low flow was less than 24 ft³/s. The mean daily 7-day low flow spanned a total of 290 days, ranging from 19.3 to 23.8 ft³/s. The minimum of daily mean values for each day spanning water years 1999–2010 was less than 24 ft³/s for 198 of the 366 days (fig. 32; table 8), which indicates that streamflow can decrease to less than 24 ft³/s during much of the year, with the normal exception of mid-January through mid-May.

    Scenarios 1-3 demonstrated that surface-water and groundwater use have an impact on Chamokane Creek, and that the groundwater and surface-water systems are in connection. However, the rates of groundwater and surface-water withdrawals, with the exception of the two hatcheries, are relatively small and would be very difficult to measure. A USGS streamflow measurement of 24 ft³/s rated as "good" indicates the measurement is within 5 percent of the true value (Rantz, 1982), or ± 1.2 ft³/s. Differences in mean annual streamflow from scenarios 1–3 typically are an order of magnitude less than that.

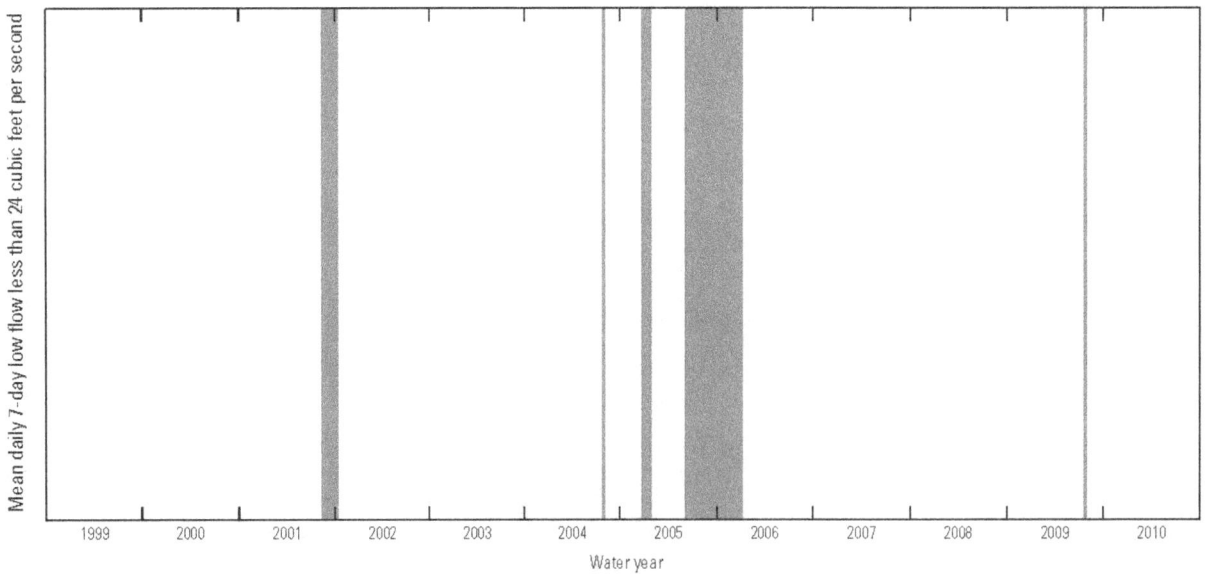

Figure 31.   Periods when mean daily 7-day low flow became less than 24 cubic feet per second for U.S. Geological Survey streamflow-gaging station 12433200, Chamokane Creek below Falls, near Long Lake, Washington, water years 1999–2010.

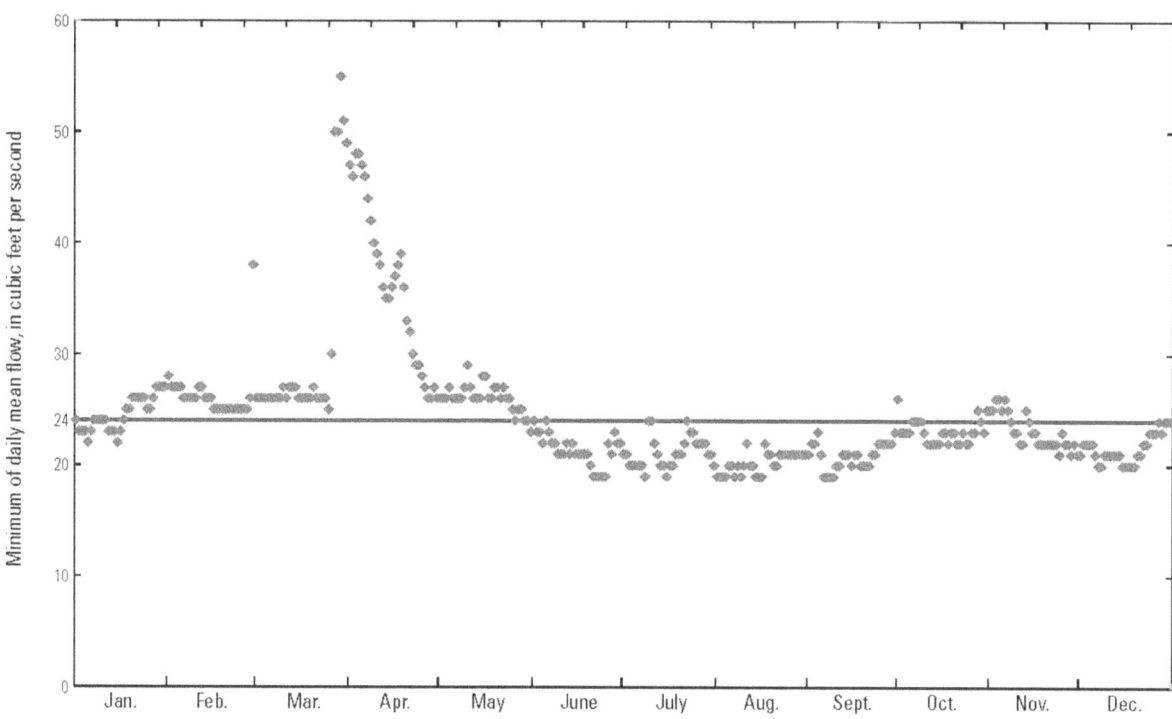

Figure 32.   Minimum of daily mean values for each day for U.S. Geological Survey streamflow-gaging station 12433200, Chamokane Creek below Falls, near Long Lake, Washington, water years 1999–2010.

In order to examine the ability to regulate streamflow due to water use in the Chamokane Creek basin, the calibrated model was used to simulate the 12-year period (water years 1999–2010) (1) with no groundwater pumping and diversions other than those associated with the hatchery operations (Scenario 4A), and (2) with twice the existing groundwater pumping and diversions at all locations other than the hatchery operations (Scenario 4B). This approach decreases (Scenario 4A) and increases (Scenario 4B) simulated water use by 0.96 ft³/s (695 acre-ft, table 7) and removes the effects of hatchery operations on streamflow, which can overwhelm all other water-use categories.

Simulated mean annual streamflow increased by 0.81 ft³/s (table 7) when the model was operated with no water use except existing hatchery operations (Scenario 4A) and decreased by 0.81 ft³/s (table 7) when the model was operated

with twice the groundwater and surface-water withdrawals at all withdrawal locations except the hatcheries (Scenario 4B). The elimination of pumping that was simulated in Scenario 4A over the 12-year simulation period would have resulted in a 7-day low flow at the gage of less than 24 ft³/s occurring on 52 fewer days than actually occurred, whereas the doubling of pumping that was simulated in scenario 4B over the simulation period would have resulted in a 7-day low flow at the gage of less than 24 ft³/s occurring on an additional 63 days than actually occurred. The cumulative monthly mean differences in streamflow are shown in figure 33. Although these differences are similar in quantity and display some of the same patterns over time, they are not exact inverses of one another, showing the system response to eliminating water use and doubling water use is not completely linear.

Table 8.   Minimum of daily mean values for each day for U.S. Geological Survey streamflow-gaging station 12433200, Chamokane Creek below Falls, near Long Lake, Washington, water years 1999–2010.

Day	Daily mean streamflow, in cubic feet per second											
	Jan.	Feb.	Mar.	Apr.	May	June	July	Aug.	Sept.	Oct.	Nov.	Dec.
1	24	28	26	47	26	24	21	19	21	26	25	21
2	23	27	26	46	26	23	21	19	22	23	25	22
3	23	27	26	48	26	23	20	19	22	23	26	22
4	23	27	26	48	27	22	20	19	23	23	26	22
5	22	27	26	47	26	24	20	20	21	23	25	22
6	23	26	26	46	26	23	20	20	19	24	26	21
7	24	26	26	44	26	22	20	19	19	24	25	20
8	24	26	26	42	26	22	19	20	19	24	24	20
9	24	26	26	40	27	21	24	19	19	24	23	21
10	24	26	27	39	29	21	24	20	20	23	23	21
11	24	27	26	38	27	21	22	22	20	22	22	21
12	23	27	27	36	26	22	21	20	21	22	22	21
13	23	26	27	35	26	21	20	20	21	22	25	21
14	23	26	27	35	26	22	20	19	21	22	24	21
15	22	26	26	36	28	21	19	19	20	22	23	20
16	23	25	26	37	28	21	20	19	21	23	23	20
17	24	25	26	38	26	21	20	22	21	23	22	20
18	25	25	26	39	26	21	21	21	20	22	22	20
19	25	25	26	36	27	21	21	21	20	23	22	20
20	26	25	27	33	27	20	21	20	20	23	22	21
21	26	25	26	32	26	19	22	20	20	22	22	21
22	26	25	26	30	27	19	24	21	21	22	22	22
23	26	25	26	29	26	19	23	21	21	23	22	22
24	26	25	26	29	26	19	23	21	22	22	21	23
25	25	25	25	28	25	19	22	21	22	22	23	23
26	25	25	30	27	24	22	22	21	22	23	22	23
27	26	25	50	26	25	21	22	21	22	23	22	24
28	27	26	50	26	25	23	22	21	22	25	21	23
29	27	38	55	27	24	22	21	21	22	24	22	24
30	27		51	26	24	22	21	21	23	23	21	24
31	27		49		23		20	21		25		24

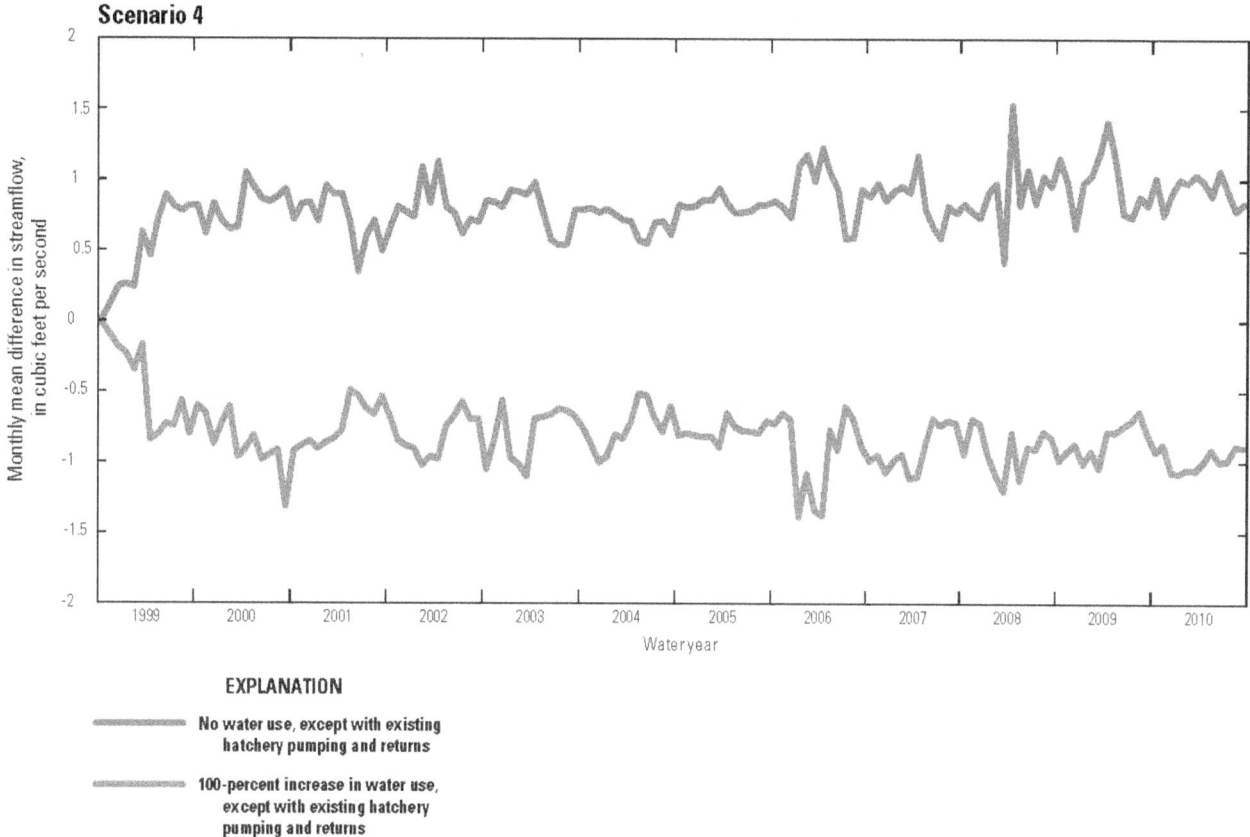

Figure 33. Difference in simulated monthly mean streamflow between existing conditions and no water use, except with existing hatchery pumping and returns, and 100-percent increase in water use, except with existing hatchery pumping and returns, Chamokane Creek basin, Washington.

Scenario 5 – Levels of Domestic and Stockwater Use That Can Be Regulated

5.   Factual Question: Is there a level of domestic or stockwater use that is too small or difficult to regulate? If so, what is that level?

Approach: The USGS will not determine if domestic or stockwater use should be regulated, but will quantify the simulated impact of domestic or stockwater use. Permit-exempt (domestic) pumping and stockwatering from both groundwater pumping and surface-water diversions were eliminated and the difference in streamflow between the base case and scenario results is the simulated effect of permit-exempt wells and stockwatering on streamflow.

The ability to regulate domestic and stockwater use in response to 7-day low flows is related to the total impact of those water-use categories on measured streamflow. To address this question, the calibrated model was used to simulate the 12-year period (water years 1999–2010)

without domestic or stockwater uses. The mean annual domestic groundwater pumping was 0.02 ft³/s (14 acre-ft; table 7, Scenario 5A). The mean annual stockwater groundwater and surface-water withdrawals were 0.01 ft³/s (7 acre-ft; table 7) and 0.02 ft³/s (14 acre-ft; table 7, Scenario 5B), respectively. Stockwatering included both groundwater pumping and surface-water diversions and domestic use included septic recharge and non-consumptive outdoor use. For this scenario, the calibrated model was run and the resulting simulated streamflows were then used as the base case for assessing the relation between stockwater and permit-exempt uses with surface-water resources. The cumulative monthly mean difference in streamflows for the two water-use categories is shown in figures 34A-B. The mean annual difference in streamflow between the calibrated model and existing conditions without permit-exempt domestic pumping (Scenario 5A) was 0.02 ft³/s (table 7). The mean annual difference in streamflow between the calibrated model and existing conditions without stockwater use (Scenario 5B) also was 0.02 ft³/s (table 7).

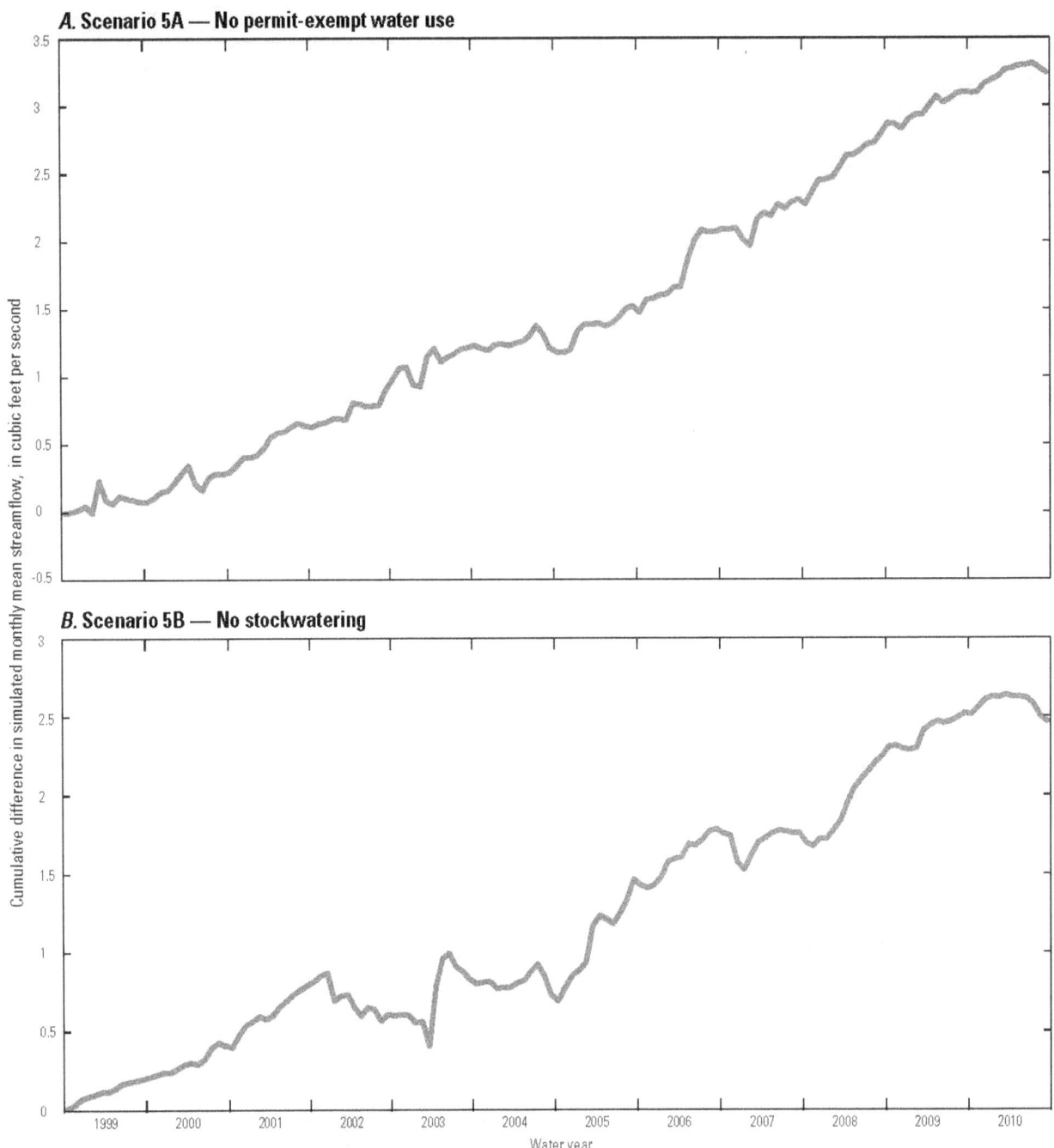

Figure 34.   Cumulative difference in simulated monthly mean streamflow between existing conditions and existing conditions without (*A*) permit exempt pumping and (*B*) stockwatering, Chamokane Creek basin, Washington.

As noted with the previous scenarios, a USGS streamflow measurement at the low-flow rate of 24 ft^3/s rated as "good" indicates the measurement is within 5 percent of the true value (Rantz, 1982), or ± 1.2 ft^3/s. Differences in mean annual streamflow of 0.02 ft^3/s simulated in scenario 5 are 1.6 orders of magnitude less than that. Therefore, the results are calculable but not measurable.

Summary of Streamflow Changes for the Five Scenarios

Another way to assess the results of the five model scenarios is to consider the change in streamflow as a percentage of the total change in model stress. For each of the scenario approaches, different categories or locations of groundwater pumping and streamflow diversions and returns were eliminated or increased, and the resulting change in streamflow was reported. In most cases, the scenario result is small because the total change in model stress (pumping and diversions) is relatively small. When viewed as a percentage of change in stress, however, an additional understanding of the connection between the scenario result and the surface-water features emerge. In Scenario 1, a larger portion of the upper-basin water use occurs in the less permeable bedrock unit with less direct connection to the mainstem of Chamokane Creek than the other aquifers. For these reasons, change in streamflow is 65 percent of the change in groundwater pumping during the 12-year simulation period (table 9). Most of the remaining 35 percent of pumping and diversions comes from groundwater storage; as the contribution of groundwater storage to pumping decreases with time, the contribution from streamflow would continue to increase with time. The elimination of all middle basin water use, including hatchery pumping and returns (Scenario 2A), results in streamflow increases that are only 30 percent of the total change in model inflows and outflows during the 12-year simulation period (table 9). This relatively low

percentage is due to the overwhelming effects of hatchery operations. Again, hatchery operations were simulated with equal rates (3.08 ft³/s) of deep groundwater pumping from the Lower aquifer and instantaneous return flows to Chamokane Creek. If middle basin water use is eliminated but hatchery operations remain at existing conditions (Scenario 2B), change in streamflow is 83 percent of the change in model stress during the 12-year simulation period (table 9). The hatcheries pump groundwater and add it to the streams, buffering the effects of other water uses in the middle basin. Scenarios 3, 4A, and 4B yield similar results, ranging from 83 to 85 percent of the change in model stress during the 12-year simulation period. Most of the non-hatchery water use occurs in the middle basin in close proximity to Chamokane Creek. Change in streamflow from the elimination of permit-exempt, or domestic, water use (Scenario 5A) is the highest percentage of change in model stress during the 12-year simulation period (97 percent; table 9). All domestic water use was simulated as groundwater pumping (no surface-water diversions) and assigned to the shallowest saturated model layer that could sustain the withdrawal. These wells were simulated to be in close connection to surface-water features, which explains the 97 percent of the eliminated pumping that results in streamflow increase during the 12-year simulation period. Stockwatering was simulated as both groundwater pumping and surface-water diversions and was located in bedrock units and permeable sediments. The change in streamflow as a percentage of change in model stress resulting from Scenario 5B is lower (61 percent) than all other scenarios except Scenario 1.

Table 9.   Simulated change in mean annual streamflow as a percentage of change in mean annual groundwater pumping and surface-water diversions and returns, Chamokane Creek basin, Washington.

[Negative values represent a decrease from base case. Positive values represent an increase from base case. **Abbreviation:** ft³/s, cubic feet per second]

Scenario No.	Change in pumping + diversions – returns (ft³/s)	Change in streamflow (ft³/s)	Change in streamflow as a percentage of change in stress (percent)
Scenario 1 - 100 percent increase in upper basin groundwater pumping and associated septic returns	0.08	−0.05	65
Scenario 2A - No middle basin groundwater or surface-water withdrawals	−0.95	0.28	30
Scenario 2B - No middle basin groundwater and surface-water withdrawals except existing hatchery operations	−0.95	0.79	83
Scenario 3 - No claims registry or permit exempt water use	−0.03	0.02	85
Scenario 4A - No groundwater pumping and surface-water diversions for all water-use categories except hatchery operations	−0.98	0.81	83
Scenario 4B -100 percent increase in groundwater pumping and surface-water diversions for all water-use categories except existing hatchery operations	0.98	-0.81	83
Scenario 5A - No permit-exempt pumping and stockwatering from groundwater pumping and surface-water diversions	−0.02	0.02	97
Scenario 5B - No stockwatering	−0.03	0.02	61

Summary

Chamokane Creek basin is a 179 mi^2 area that borders and partially overlaps the Spokane Indian Reservation in southern Stevens County in northeastern Washington. Aquifers in the Chamokane Creek basin are part of a sequence of glaciofluvial and glaciolacustrine sediments that may reach a total thickness of about 600 feet. In 1979, most of the water rights in the Chamokane Creek basin were adjudicated by the United States District Court requiring regulation in favor of the Spokane Tribe of Indian's senior water right. A court-appointed Water Master regulates junior water rights when the mean daily 7-day low flow becomes less than 24 cubic feet per second (ft^3/s) (27 ft^3/s for rights issued after December 1988) at Chamokane Falls, as recorded at U.S. Geological Survey (USGS) streamflow-gaging station 12433200; regulation has been necessary in 3 recent years (2001, 2005, and 2009). Additionally, the non-Reservation areas of the basin are closed to additional groundwater or surface-water appropriations, with the exception of permit exempt uses of groundwater.

A three-dimensional, transient numerical model of groundwater and surface-water flow was constructed for the Chamokane Creek basin aquifer system to better understand the groundwater-flow system and its relation to surface-water resources. The model described in this report can be used as a tool by water-management agencies and other stakeholders to quantitatively evaluate the effects of potential increases or decreases in groundwater pumping on groundwater and surface-water resources within the basin.

The Chamokane Creek model was constructed using the U.S. Geological Survey (USGS) integrated model, GSFLOW. GSFLOW was developed to simulate coupled groundwater and surface-water resources. The model uses 1,000-foot grid cells that subdivide the model domain by 102 rows and 106 columns. Six hydrogeologic units in the model are included in eight model layers. Chamokane Creek and its major tributaries are included in the model as streamflow-routing cells. Daily precipitation and temperature were specified in the model, and groundwater recharge was computed by GSFLOW. Groundwater pumpage and surface-water diversions and returns specified in the model were derived from monthly and annual pumpage values previously estimated from another component of this study and new data reported by study partners.

The model simulation period is water years 1980–2010 (October 1, 1979, to September 30, 2010), but the model was calibrated to the transient conditions for water years 1999–2010 (October 1, 1998, to September 30, 2010). Calibration was completed by using traditional trial-and-error methods and automated parameter-estimation techniques. The model adequately reproduces the time-series groundwater-level hydrographs and daily streamflow. At well observation points, the average difference between simulated and measured hydraulic heads is 7 feet with a root-mean-square error divided by the total difference in water levels of 4.7 percent. Simulated river streamflow was compared to measured streamflow at the USGS streamflow-gaging station. Annual differences between measured and simulated streamflow for the site ranged from -63 to 22 percent. Calibrated model output includes a 31-year estimate of monthly water budget components for the hydrologic system.

Five applications (scenarios) of the model were completed to obtain a better understanding of the relation between groundwater pumping and surface-water resources and groundwater levels. The calibrated transient model was used to evaluate: (1) the connection of the upper and middle basin groundwater systems, (2) the effect of surface-water and groundwater uses in the middle basin, (3) the cumulative impacts of claims registry and permit-exempt wells on Chamokane Creek streamflow, (4) the frequency of regulation due to impacted streamflow, and (5) the levels of domestic and stockwater use that can be regulated. The simulation results indicated that streamflow is affected by the existing pumpage from 1999 through 2010. The mean annual difference in streamflow between the calibrated model and the model scenarios exhibited similarities in the absolute amount, yet patterns in monthly mean differences varied from one scenario to the next. The change in streamflow that resulted from each scenario generally was inversely related to the total change in pumpage eliminated or increased in the model scenarios.

For Scenario 1, the simulated effect a 100 percent increase in upper basin groundwater pumping indicates the mean annual difference in streamflow at Chamokane Creek below Falls, near Long Lake, Washington, to be a decrease of about 0.05 ft^3/s.

The mean annual difference in streamflow between the calibrated model and Scenario 2A (no middle basin water use) streamflows was an increase of about 0.3 ft^3/s. To separate the effects of the large groundwater pumping and streamflow returns associated with hatchery operations, Scenario 2B was simulated with no middle basin water use except for existing hatchery operations. The difference in simulated mean annual streamflow for this scenario was about 0.8 ft^3/s.

For Scenario 3, the simulated effect of no claims and permit-exempt water use indicates the mean annual difference in streamflow at Chamokane Creek below Falls, near Long Lake, Washington, to be an increase of about 0.02 ft^3/s.

In order to examine the ability to regulate streamflow due to water use in the Chamokane Creek basin, the calibrated model was used to simulate water years 1999–2010 (1) without groundwater pumping and diversions, except hatchery operations (pumping and return flows) (Scenario 4A), and (2) with twice the existing groundwater pumping and diversions, except existing hatchery operations (Scenario 4B). Simulated mean annual streamflow increased by 0.8 ft^3/s when the model was operated with no water use except existing hatchery operations. Simulated mean annual streamflow decreased by 0.8 ft^3/s when the model was operated with twice the groundwater and surface-water withdrawals (excluding hatchery operations).

To address the question of "Is there a level of domestic or stockwater use that is too small or difficult to regulate," the model was operated without domestic and stockwater use and the difference in streamflow was compared to the measurement accuracy at the USGS streamflow-gaging station. The mean annual difference in streamflow between the calibrated model and existing conditions without permit-exempt domestic withdrawals (Scenario 5A) was an increase of about 0.02 ft³/s. The mean annual difference in streamflow between the calibrated model and existing conditions without stockwater use (Scenario 5B) also was an increase of about 0.02 ft³/s. This difference is about 2 percent of the ± 1.2 ft³/s error associated with a "good" USGS streamflow measurement of 24 ft³/s.

Acknowledgments

The authors of this report gratefully acknowledge the landowners who shared their knowledge and concerns about water resources in the Chamokane Creek basin and allowed access to their property for data collection. Without their consent, it would not have been possible to collect most of the data required for this study. Brian Crossley, Spokane Tribe Natural Resources, responded to numerous inquiries for data, references, and contact information pertaining to water resources on the Spokane Indian Reservation. Charlie Kessler, Stevens County Conservation District, provided insights on the surface-water system of the basin as well as detailed information collected as part of his Chamokane Creek Watershed Management work. James Lylera, Chamokane Creek Water Master, provided long-term water-level data and water-use information for the basin.

References Cited

Anderman, E.R., and Hill, M.C., 2000, MODFLOW-2000, the U.S. Geological Survey Modular Groundwater Model-Documentation of the Hydrogeologic-Unit Flow (HUF) Package: U.S. Geological Survey Open-File Report 2000-342, 89 p. (Also available at http://pubs.er.usgs.gov/publication/ofr00342.)

Anderson, M.R., and Woessner, W.W., 1992, Applied groundwater modeling simulation of flow and advective transport: San Diego/New York/Boston/London/Sydney/Tokyo/Toronto, Academic Press, Inc., 381 p.

Atwater, B.F., 1986, Pleistocene glacial-lake deposits of the Sanpoil River Valley, northeastern Washington: U.S. Geological Survey Bulletin 1661, 39 p. (Also available at http://pubs.er.usgs.gov/publication/b1661.)

Buchanan, J.P., Wozniewicz, J.V., and Lambeth, R.H., 1988, Hydrogeology of the Chamokane Valley aquifer system: Upper Columbia United Tribes Fisheries Center, Eastern Washington University, Fisheries Technical Report No. 20, 69 p.

Carrara, P.E., Kiver, E.P., and Stradling, D.F., 1995, Surficial geologic map of the Chewelah 30' × 60' quadrangle, Washington and Idaho: U.S. Geological Survey Miscellaneous Investigations Series Map I-2472, 1 sheet, scale 1:100,000.

Carrara, P.E., Kiver, E.P., and Stradling, D.F., 1996, The southern limit of Cordilleran ice in the Colville and Pend Oreille valleys of northeastern Washington during the late Wisconsin glaciation: Canadian Journal of Earth Sciences, v. 33, no. 5, p. 769–778.

Conlon, T., Lee, K., and Risley, J., 2003, Heat tracing in streams in the central Willamette Basin, Oregon, in Stonestrom, D.A., and Constantz, J., eds., Heat as a tool for studying the movement of ground water near streams: U.S. Geological Survey Circular 1260, chapter 5, p. 29-34. (Also available at http://pubs.usgs.gov/circ/2003/circ1260/.)

Criss, R.E., and Winston, W.E., 2008, Do Nash values have value? Discussion and alternate proposals: Hydrological Processes, v. 22, p. 2723-2725.

Daly, Chris, Neilson, R.P., and Phillips, D.L., 1994, A statistical-topographic model for mapping climatological precipitation over mountainous terrain: Journal of Applied Meteorology, v. 33, p. 140-158.

Daly, C., Taylor, G.H., and Gibson, W.P., 1997, The PRISM approach to mapping precipitation and temperature: in Reprints: 10th Conference on Applied Climatology, Reno, American Meteorological Society, p. 10-12.

Doherty, J.E., 2010, PEST, Model-independent parameter estimation—User manual (5th ed., with additions): Brisbane, Australia, Watermark Numerical Computing.

Doherty, J.E., and Hunt, R.J., 2010, Approaches to highly parameterized inversion—A guide to using PEST for groundwater-model calibration: U.S. Geological Survey Scientific Investigations Report 2010–5169, 59 p. (Also available at http://pubs.usgs.gov/sir/2010/5169/.)

Drost, B.W., Ely, D.M., and Lum, W.E., 1999, Conceptual model and numerical simulation of the ground-water flow system in the unconsolidated sediments of Thurston County, Washington: U.S. Geological Survey Water-Resources Investigations Report 99-4165, 106 p. (Also available at http://pubs.usgs.gov/wri/wri994165/.)

Dunne, T., and Black, R.G., 1970, An experimental investigation of runoff production in permeable soils: Water Resources Research, v. 6, p. 478-490.

Ely, D.M., Bachmann, M.P., and Vaccaro, J.J., 2011, Numerical simulation of groundwater flow for the Yakima River basin aquifer system, Washington: U.S. Geological Survey Scientific Investigations Report 2011-5155, 90 p.

Ely, D.M., and Kahle, S.C., 2004, Conceptual model and numerical simulation of the groundwater flow system in the unconsolidated deposits of the Colville River Watershed, Stevens County, Washington: U.S. Geological Survey Scientific Investigations Report 2004-5237, 72 p.

Freeze, R.A., and Cherry, J.A., 1979, Groundwater: Englewood Cliffs, N.J., Prentice-Hall, 604 p.

Golder Associates, Inc., 2008, Final report, conceptual site models for Martha Boardman/Kokanee Meadows and Ford/Newhouse Lane Water Systems: Coeur d'Alene, Idaho, prepared for Indian Health Service, Spokane, Wash., May 14, 2008, variously paginated.

Harbaugh, A.W., 2005, MODFLOW-2005, the U.S. Geological Survey modular groundwater model-the Groundwater Flow Process: U.S. Geological Survey Techniques and Methods 6-A16, variously paginated.

Hay, L.E., Wilby, R.L., and Leavesly, G.H., 2000, A comparison of delta change and downscaled GCM scenarios for three mountainous basins in the United States: Journal of the American Water Resources Assoc., v. 36, no. 2, p. 387-397.

Hewlett, J.D., and Hibbert, A.R., 1967, Factors affecting the response of small watersheds to precipitation in humid regions: Forest Hydrology, Pergamon Press, Oxford, p. 275-290.

Hewlett, J.D., and Nutter, W.L., 1970, The varying source area of streamflow from upland basins: Proceedings of the Symposium on Interdisciplinary Aspects of Watershed Management, Bozeman, Montana, p. 65-83.

Hill, M.C., 1998, Methods and Guidelines for effective model calibration: U.S. Geological Survey Water-Resources Investigations Report 98-4005, 90 p.

Hill, M.C., and Tiedeman, C.R, 2003, Weighting observations in the context of calibrating ground-water models, *in* Kovar, K., and Zbynek, H., eds., Calibration and reliability in groundwater modeling, a few steps closer to reality: International Association of Hydrological Sciences Publication, v. 277, p. 196–203.

Hill, M.C., and Tiedeman, C.R., 2007, Effective groundwater model calibration, with analysis of sensitivities, predictions, and uncertainty: New York, New York, Wiley, 455 p.

Horton, R.E., 1933, The role of infiltration in the hydrological cycle: American Geophysical Union Transactions, v. 23, p. 479-482.

Jensen, M.E., and Haise, H.R., 1963, Estimating evapotranspiration from solar radiation: Proceedings of the American Society of Civil Engineers, Journal of Irrigation and Drainage, v. 89, no. IR4, p. 15-41.

Kahle, S.C., Longpré, C.I., Smith, R.R., Sumioka, S.S., Watkins, A.M., and Kresch, D.L., 2003, Water resources of the groundwater system in the unconsolidated deposits of the Colville River watershed, Stevens County, Washington: U.S. Geological Survey Water-Resources Investigations Report 03-4128, 76 p.

Kahle, S.C., Taylor, W.A., Lin, Sonja, Sumioka, S.S., and Olsen, T.D., 2010, Hydrogeologic framework, groundwater and surface-water systems, land use, pumpage, and water budget of the Chamokane Creek basin, Stevens County, Washington: U.S. Geological Survey Scientific Investigations Report 2010-5165, 60 p.

Kessler, C., 2008, Chamokane Creek Watershed Needs Assessment: Colville, Wash., Stevens County Conservation District, 68 p.

Kiver, E.P., and Stradling, D.F., 1982, Quaternary geology of the Spokane area, in Roberts, S., and Fountain, D., eds., 1980 Field Conference Guidebook: Spokane, Wash., Tobacco Root Geological Society, p. 26-44.

Leavesley, G.H., Lichty, R.W., Troutman, B.M., and Saindon, L.G., 1983, Precipitation-runoff modeling system—User's manual: U.S. Geological Survey Water-Resources Investigations Report 83-4238, 207 p.

Leavesley, G.H., Markstrom, S.L., Viger, R.J., and Hay, L.E., 2005, USGS Modular Modeling System (MMS) — Precipitation-Runoff Modeling System (PRMS) MMS-PRMS, *in* Singh, V., and Frevert, D., eds., Watershed Models: Boca Raton, Fla., CRC Press, p. 159-177.

Leavesley, G.H., Restrepo, P.J., Markstrom, S.L., Dixon, M., and Stannard, L.G., 1996, The modular modeling system (MMS)-User's manual: U.S. Geological Survey Open-File Report 96-151, 200 p.

Magirl, C.S., and Olsen, T.D., 2009, Navigability potential of Washington rivers and streams determined with hydraulic geometry and a geographic information system: U.S. Geological Survey Scientific Investigations Report 2009–5122, 22 p. (Also available at http://pubs.usgs.gov/sir/2009/5122/.)

Markstrom, S.L., Niswonger, R.G., Regan, R.S., Prudic, D.E., and Barlow, P.M., 2008, GSFLOW-Coupled Groundwater and Surface-water FLOW model based on the integration of the Precipitation-Runoff Modeling System (PRMS) and the Modular Groundwater Flow Model (MODFLOW-2005): U.S. Geological Survey Techniques and Methods 6-D1, 240 p., accessed March 23, 2010, at http://water.usgs.gov/nrp/gwsoftware/gsflow/gsflow.html.

McLucas, G.B., 1980, Surficial geology of the Springdale and Forest Center quadrangles, Stevens County, Washington: Washington Division of Geology and Earth Resources Open-File Report 80-3, 29 p., 2 pls.

Molenaar, Dee, 1988, The Spokane aquifer, Washington—Its geologic origin and water-bearing and water-quality characteristics: U.S. Geological Survey Water-Supply Paper 2265, 74 p.

Multi-Resolution Land Characteristics Consortium, 2008, NLCD 2001 Land Cover Class Definitions, National Land Cover Database, accessed May 27, 2010, at http://www.mrlc.gov/nlcd_definitions.php.

Niswonger, R.G., Panday, Sorab, and Ibaraki, Motomu, 2011, MODFLOW-NWT, A Newton formulation for MODFLOW-2005: U.S. Geological Survey Techniques and Methods 6–A37, 44 p.

Niswonger, R.G., and Prudic, D.E., 2005, Documentation of the Streamflow-Routing (SFR2) Package to include unsaturated flow beneath streams—A modification to SFR1: U.S. Geological Survey Techniques and Methods, Book 6, Chap. A13, 47 p.

Niswonger, R.G., Prudic, D.E., and Regan, R.S., 2006, Documentation of the Unsaturated-Zone Flow (UZF1) Package for modeling unsaturated flow between the land surface and the water table with MODFLOW-2005: U.S. Geological Survey Techniques and Methods 6-A19, 62 p.

Powell, D.S., Faulkner, J.L., Darr, D., Zhu, Z., MacCleery, D.W., 1998, Forest resources of the United States, 1992: Fort Collins, Colo., General Technical Report, U.S. Department of Agriculture, Forest Service, Rocky Mountain Forest and Range Experiment Station, 132 p.

Rantz, S.E., 1982, Measurement and computation of streamflow, volume 1—Measurement of stage and discharge: U.S. Geological Survey Water-Supply Paper 2175, 284 p.Richmond, G.M., Fryxell, R., Neff, G.E., and Weis, P.L., 1965, The Cordilleran ice sheet of the northern Rocky Mountains, and the related Quaternary history of the Columbia Plateau, in Wright, H.E., Jr., and Frey, D.G., eds., The Quaternary of the United States: Princeton, N.J., Princeton University Press, p. 231-242.

Rittenhouse-Zeman and Associates, Inc., 1989, Well construction and aquifer testing, Deep aquifer at Galbraith Springs, Stevens County, Washington: Beaverton, Oreg., Rittenhouse-Zeman and Associates [variously paged].

Stoffel, K.L., Joseph, N.L., Waggoner, S.Z., Gulick, C.W., Korosec, M.A., and Bunning, B.B., 1991, Geologic map of Washington northeast quadrant: Washington Division of Geology and Earth Resources, Geologic map GM-39, scale 1:250,000.

Troendle, C.A., 1985, Variable source area models, in Anderson, M.G., and Burt, T.P., eds., Hydrological Forecasting: John Wiley and Sons Ltd., p. 347-403.

Troutman, B.M., 1985, Errors and parameter estimation in precipitation-runoff modeling—2. Case Study: Water Resources Research, v. 21, no. 8, p. 1,214-1,222.

United States v. Anderson, United States District Court, 23 Jul. 1979.

United States v. Anderson, United States District Court, 11 Aug. 2006.

U.S. Department of Agriculture, 1994, State Soil Geographic (STATSGO) database-Data use information: Fort Worth, Texas, Soils Conservation Service, National Cartography and GIS Center.

Vaccaro, J.J., and Olsen, T.D., 2007, Estimates of ground-water recharge to the Yakima River Basin Aquifer System, Washington, for predevelopment and current land-use and land-cover conditions: U.S. Geological Survey Scientific Investigations Report 2007–5007, 30 p. (Also available at http://pubs.water.usgs.gov/sir2007/5007.)

Viger, R.J., and Leavesley, G.H., 2007, The GIS Weasel User's Manual: U.S. Geological Survey Techniques and Methods 6-B4, 201 p.

Waitt, R.B., Jr., and Thorson R.M., 1983, The Cordilleran ice sheet in Washington, Idaho, and Montana, in Wright, H.E., and Porter, S.C., eds., Late-Quaternary environments of the United States, v. 1: Minneapolis, University of Minnesota Press, p. 53-70.

Washington State Department of Ecology, 2003, Well logs: accessed May 18, 2012, at http://apps.ecy.wa.gov/welllog/.

Washington State Department of Ecology, 2006, Water Resources Program – Water Rights, accessed August 22, 2006, at http://www.ecy.wa.gov/programs/wr/rights/water-right-home.html.

Washington State Department of Health, Division of Radiation Protection, 1991, Final environmental impact statement—Closure of the Dawn Mining Company uranium millsite in Ford, Washington: Olympia, Washington State Department of Health, 3 vol., variously paged.

Western Regional Climate Center, 2010, Western U.S. climate historical summary for Wellpinit, Washington: Western Regional Climate Center database, accessed January 6, 2010, at http://www.wrcc.dri.edu/cgi-bin/cliMAIN.pl?wa9058.

Whiteman, K.J., Vaccaro, J.J., Gonthier, J.B., and Bauer, H.H., 1994, The hydrogeologic framework and geochemistry of the Columbia Plateau aquifer system, Washington, Oregon, and Idaho: U.S. Geological Survey Professional Paper 1413-B, 73 p.

Willis, B., 1887, Changes in river courses in Washington Territory due to glaciation: U.S. Geological Survey Bulletin 40, p. 477-480.

Wozniewicz, J.V., 1989, Hydrogeology of the Chamokane valley aquifer system: Cheney, Eastern Washington University, Master of Science thesis, 172 p.

Zahner, R., 1967, Refinement in empirical functions for realistic soil-moisture regimes under forest cover, *in* Sopper, W.E., and Lull, H.W., eds., International Symposium of Forest Hydrology: New York, Pergamon Press, p. 261-274.

Zhu, Z., and Evans, L.D., 1992, Mapping midsouth forest distributions: Journal of Forestry, v. 90, no. 12, p. 27-30.

www.ingramcontent.com/pod-product-compliance
Lightning Source LLC
Chambersburg PA
CBHW081557170526
45166CB00009B/2727